We stand on the edge of things, gazing into an abyss, that, quite literally, gazes also into us.

We have been here before, though, standing on this same edge of things, gazing into the same abyss. Perhaps, when you really look at it, get beyond the surface, it's also been the same gaze.

We were here when we stopped scavenging wild fruits, nuts, seeds, and the kills of other animals, and formed small, loosely co-operative hunting bands.

We were here when, rather than following our preferred form of steak over the plains, or through the jungle, or up mountains, we began to bring our small, loosely-co-operative hunting bands together into larger groups, groups which included women and children, and which served functions beyond the finding and preparing of food.

We were here when the first machines revolutionised the way we performed the tasks of our everyday lives, when we knew, for the first time, what it meant to be able to take brief breaks from the struggle to simply stay alive, to have "leisure time."

And now, we're here again, at another point in the evolution of our aberrant species, standing on the edge of the "modern" world, gazing into the futuristic, virtual-reality abyss of the "modem" world, watching its flashing lights and yelping sirens, struggling to interpret the graphics of its lexicon – feeling the abyss no longer just gazing back, but actively drawing us down, pulling us into its unknown, seemingly endless, embrace.

In the modern world, we harnessed the power of machines to produce things that had previously been unimaginable, in quantities and within a timescale that, before, we could only have dreamed of. The modern world saw us travelling, in large numbers, between communities and societies, countries and continents. It raised the standard of living, increased life expectancy, educational achievement, and employment prospects. In the modern world, we climbed Maslow's Pyramid, pulling ourselves up from an all-consuming focus on safety and biological needs – where our next meal was coming from, how we would keep warm through the winter, how we could best ensure our children would survive, to an age and standard of health where they, too, would be useful, would be able to carry out the business of the tribe, first into a full engagement in becoming and being social animals – establishing the rituals and conventions that bind societies together, the laws that keep them relatively stable, and the personal connections that are, fundamentally, the reason everything else works, then, towards the end of the modern age, throughout the twentieth century, into the upper strata of Maslow's hierarchy, grasping at the closest fruits of a sense of self, an idea, and the realisation of that idea, of "I" instead of "we". In the modern age, we achieved more, had more, did more, *were* more, than our ancestors could ever have imagined possible. We began to build bridges from cables, data, and thoughts, and laughed at the idea that bridges had to be solid, made of stone or steel or wood, laughed, as our bridges changed their form, function, and span, at those who believed that bridges had to be enduring, eternal structures. Towards the end of the modern age, nothing was forever. At the end of the modern world, we found ourselves standing with processes that meant we could produce vast quantities of whatever we wanted at next to no cost, machines that meant labour and travel took very little time, and almost no energy, at the edge of an abyss a million miles away from what we had once been – frail, vulnerable, frightened animals, bereft of effective natural weapons, naked of natural coverings and natural defences against the elements, who were yet capable of outwitting and defeating things bigger, stronger, and faster than ourselves, animals without fur that could yet survive winters harsh enough to kill many better-coated beasts, animals capable of adapting to change so fully and completely that no shift in the natural world could destroy our species, creatures that had risen from frail, naked insignificance to become powerful apex predators.

Human beings evolved and advanced, always, by some form of stealing fire from the gods. Time and again, we took the fire – both literal and metaphorical – that we needed without recompense or thanks, because, we reasoned, the gods were only there because we had created them; what was theirs was, by default, ours.

First, we made gods that we could tame, and order to save, protect, and aid us – statues and icons that we would place in a suitable part of our home, who would, in exchange for small tokens of faith and devotion, provide us with wealth, good fortune, power, health, fertility, or whatever else we demanded.

Then, tiring of making these tame gods with their feet of clay, we began making gods of ourselves, enhancing our beauty, increasing our personal wealth, promoting "BRAND ME!!!" at every opportunity, and, like all gods, demanding suitable tribute in exchange for the possibility of later reward.

Eventually, we got tired with the demands of being gods, and began creating god-like machines that would take over these quotidian frustrations, leaving us free to revel in our status, power, and wealth.

It is these creations that gaze into us, as we gaze – not for the first time, but, perhaps, for the last – into the abyss of a future that is as inevitable as it is unknown.

As human beings, we have never consciously worked towards an evolutionary endpoint, but have merely arrived through doing what we've always done, changing things a little here, coming up with an interesting new train of thought there, tinkering with the edges of the path, until – like Wile E Coyote – we suddenly find ourselves running in mid-air, a sheer drop below us. A sheer drop that, in this case, ends in the abyss of gods and machines – and with one question: which is which? Which are the gods, and which the machines?

4

GENESIS

The first fire we stole was literal – the flames that would warm us, our homes, and our food.

History does not record how many habitats, species, and lives were lost as the first humans grappled to gain mastery over the power of those first flames, but, given our enduring ability for self-destruction, and our unique habit of turning on healthy members of our own kind, for reasons that have little or nothing to do with survival or competition for resources, it seems unlikely that our contraband would have been without impact on our environment. We probably crashed and burned several times before we became the gods of this god-born force that would come to revolutionise our lives. Fire's power is, ultimately, destructive – it is only with skilled harnessing and handling that this destructive capability can be put to creative uses, only with expertise can new life be brought forth from fiery death. And, let's be honest, humans don't have much of a track record with skill or expertise.

We stole fire from the gods in order to become human. The most human endeavours and traits – creativity, love, anger, focus – all have fire and flame as their central metaphor. We are "burning with desire", we have a "white-hot focus", we "feel the heat of jealousy". Flame-haired people are said to be less temperate, less in control of themselves than those with more typical colouring. When we see a red bedroom, we smile and nod knowingly, while a red car sets alarm bells ringing. When we wish to destroy something absolutely, we burn it.

"Where one burns books, one eventually burns people", observed Brecht, a German author, during the height of the Nazi regime. And, with the provocatively named 'Kindle', Amazon have subtly conveyed that at least one of their goals is the destruction of the physical book – even though there may be no conscious awareness of this, the echoes and reverberations of the name chosen for their device have a life and a meaning of their own. Physical books will "burn" in the fires of the realisation that they are expensive, bulky, and take up too much space. And people? People will burn as the "democratisation of publishing" leads to lower and lower revenue for individual authors, and ever-greater numbers of poor quality books thrust upon unsuspecting readers.

But phoenixes rise from the ashes of every fire, and the creation that may well come about from the rise of ebook publishing is a viable platform for niche writers, a door through which those who may have previously thought reading "wasn't for them" can enter, and, perhaps, a renewal of the love of reading among children raised to believe that, if it doesn't have apps and a screen, it's "boring."

Socrates was alive when writing first began to replace the oral tradition, and he was very vocal in his belief that this "new technology" was a threat to humanity, and to the human capacity for memory. We would, he believed, become so reliant on the written word that our mental capacity for recall would atrophy – writing would, Socrates believed, eventually rob us forever of the ability to remember. Society, he claimed, would lose all cohesion, as people would no longer need to have any skill at oration, or to engage people through their presence and voice alone. People would no longer travel to hear a stirring speech, or to see a play performed, because they wouldn't need to – these things would be written down for them, and brought to them. There would be no need, so Socrates believed, for communal activities.

And yet, the exchange of a physical book – particularly one already well-loved and well-read – engenders a new level of intimacy and community; as our glance skims across the page at which a creased and battered spine causes a book to fall open, as we read underlined phrases and notes pencilled in the margins, we feel a connection with all those readers gone before us – a window is opened into the mind of another human being, someone that, possibly, we have never and will never meet, that will not be lightly closed.

Wayne Gooderham, author of 'Dedicated To', is a collector of secondhand books with handwritten dedications from the giver to the recipient – tangible proof that the written word did not, as Socrates feared it would, put an end to the time and commitment one human being could have for another.

And yet, to a certain extent, the advent of the written word *has* resulted in a loss of patience – we no longer have the patience to travel miles to hear a particular speaker, and sit for hours listening to them – we expect them to do a podcast, to have a YouTube channel, we buy their books rather than going to their conferences.

Will anyone in the future have the patience to wait until their local bookshop or library finally brings in a copy of the book they want to read? Will we even *have* bookshops and libraries, or will we have lost patience with the hit-and-miss nature of such places – the financial restrictions of the former, and the limited period of enjoyment of a thing permitted by the latter? Will future students have the patience to read set texts, and to think about what they've read, or will examinations become merely a test of how quickly students can find Wikipedia articles, and how adept they are at rephrasing the thoughts of others?

Perhaps there will no longer be schools, exams, or students – perhaps individuals will simply explore a subject or area of interest for its own sake, in their own time and fashion – watching TED talks, reading blogs, joining online debates, and publishing blogs and YouTube videos of their own on the subject.

Perhaps, through the forceful promotion of the concept of the digital book, epublishing has signed its own death warrant – perhaps, in the future, there will be no books at all, either physical or digital, but only blogs and videos. Perhaps the 'job title' of 'author' will vanish altogether, and the

'marketable product' will be advertising
space on an individual's blog, or a
blogger's appearance - in the flesh! - at
events; after all, no matter how much things
change, and however much we welcome the
change, and hail it as 'progress', there
always seems to be some sort of value to us
in the 'old fashioned' things we left
behind.
And then, of course, there is the fact that
nothing in this world is ever entirely new,
no matter what claims of 'novelty' are made
for it.
   Facebook, for example, takes us back to
prehistoric times, when cavedwellers would
write on one another's walls, alerting the
tribe to the fact that the mammoths were
migrating, or that a particular bird would
be passing over the area on its way North or
South, or, given that human behaviour
doesn't change much from one era to the
next, telling their neighbours what they
were having for dinner. We're taken back to
nature, to the world of the descendants of
dinosaurs, as we tweet, we erect firewalls -
reminiscent of the medieval battles
involving rows of flaming torches, and
burning arrows fired from castle
battlements. We give our computers
instructions in code, an ancient invention
with an illustrious pedigree, and Trojan
horses lay waste to our files and systems,
as their namesake once laid waste to a
magnificent city, and we can set Cerberus as
a guard.
   For all its much-hyped 'novelty',
technology still relies on the written word,

its systems and practices are named from antiquity, and all content still has to be physically transmitted. Even the word 'computer' is ancient, harking back to the C17th – long before 'a computer' was a human being engaged in accounting tasks, the 17th-century word being a compound of the Latin *'com'*(meaning, 'with'), and *'putare'*(meaning, 'reckoning') – the term 'to reckon' originally meant 'to reduce to a manageable quantity.'

Meanwhile, the word 'cyber' comes from the Greek *'kybernetes'*, which referred to a person or device engaged in the steering of a ship. 'Bionic', likewise, has a Greek origin – *'bios'*, or 'life'.

And then we have Bluetooth. Surely *that* can only be modern? Well, the Bluetooth symbol is comprised of the Swedish runes Hagall and Bjorken, the initials of King Harald of Denmark (circa 935AD), who was known a "Blatand" – or "Bluetooth."

And if you've ever wondered why being someone's Facebook 'friend' is such a big deal, the Old English verb *'freogan'*, from which we get our nouns friend, friendship, etc, meant "to bestow favour upon."

We connect to many possible futures through a present-day medium that is firmly rooted in the past. Computers are no fantastic leap of creative imagination, but a natural, expected point on the path of human evolution, and the evolution of human language, human communication, and human capability.

Many of us use technology as a way to leave the past behind, yet it is only by

looking at technology's roots, and following
their pathways into our present, that we
will be in a position to see where things
are headed - and to either follow, or run
ahead.

Progress is intimately connected to loss – we can't move forward without leaving something behind, and it is this awareness that informs the "scare-mongering" that accompanies the journey to any new point on the path of our evolution.

Let's address the fears of things we shall lose to technology one by one, section by section.

Memory.

The fear that an increasing reliance on computers will reduce our innate capacity for memory is nothing new – as we've already noted, Socrates was concerned that the advent of the written word would result in a catastrophic loss of memory among the population at large, as the oral tradition slowly crumbled under this "new technology."

In 2009, the authors of 'Trust Agents', Chris Brogan and Julien Smith, claimed that "human memory is slowly becoming obsolete."

Television was meant to sound the death-knell for radio, just as radio was meant to mean the end for live performance. Neither happened, and, for all the publicity about technology's impact on human memory, it would seem that, far from the attitude of "oh well, we've got computers – remembering things isn't really all that important now, is it?", people are more concerned than ever about *losing* their memory – drugs companies still continue their search for the Holy Grail of medicine, a 'cure' for Alzheimer's and dementia, books on 'brain training' still sell well, and we still worry when we "seem to be forgetting things."

Human memory will never become obsolete, because it would take too long, and cost too much, to build a computer capable of the full range and complexity of human memory.

For example, I remember that vaccinating children against measles, mumps and rubella is important, and also the furore over the claim by British doctor Andrew Wakefield that the MMR vaccine 'caused autism'. I remember, too, that, as a child, I contracted measles three days after receiving the vaccine. I remember the red Wellington boots I had as a child, and the photograph in which I am wearing them, along with red shorts and a red and white striped t-shirt, as I help my father in the garden. I remember the time I cut my shin completely open, climbing a steep embankment in that same garden, and I remember the grey corduroy of the sofa my parents laid me on while the family doctor, who was also a good friend of my parents, stitched my leg up.

I remember my friend Paul's phone number, but also that he can't usually be reached over the weekend, because of his job as a youthworker. I remember how to get to my friend Adam's house, but also that he's usually out on Monday nights, and that his wife, Shona, works shifts.

A computer can't be programmed – yet, possibly ever – with the *emotional* connections of memories, and this simple fact is what will save and preserve the unique value of *human* memory.

I remember the feel of wet sand between my toes, but I don't recall the chemical make up of sand, though I know that sand and glass share constituent properties. I don't remember the dictionary definition of sarcasm, but I remember the times I've heard it used well.

Human memory is also prone to errors, to mis-remembered details that serve to make an experience more relevant.

For example, I 'remember' that, the first time I ever met the woman who is now my wife, she was wearing a red dress and combat boots. She informs me that, in fact, she was wearing black knee-length boots, not combat boots at all. And yet my memory of her in a dress and combat boots persists, because it suits my sense of the kind of person, the kind of woman, she is - a woman who would see nothing amiss in wearing a dress with combat boots, a woman feminine yet practical.

No computer, however powerful or well-programmed, will every have a memory like this, and it is the loops and whorls, the tangents and diversions, of individual human memory that make it special, that raise us above and beyond the level of 'skin-clad machines'. It is our errors and impressions that make us human, that give us our power. In our falsehoods, we become gods.

The form and function of human memory might change - we might stop using it to remember grocery lists, for example, or when bills need to be paid, which might lead, over time, to changes in the structure of the human brain, as certain areas atrophy, and others increase due to changes in what we remember, and how we remember it, but we will never surrender the very thing that makes us gods to the machines of our creation.

I may save my shopping list to a mobile device, but it will be my memory – my human memory – that recalls the first time I tasted chorizo, and all the ways – which are nothing to do with ingredients, country, or method of manufacture – that chorizo is nothing like salami. (I was fifteen, on a school trip to the Loire Valley region of France, the youngest of a group of eight, yet, because of my size, the only one to be served in the French bars, with their lower drinking age of sixteen. We were at a medieval re-enactment festival, and I ate chorizo in paella, at a rough wooden table in a long marquee. Later, when I would see chorizo in English supermarkets, I would buy it with chicken and bacon, and serve in a smoked cheese salad. Whereas salami is coarse, and tastes like a meal given hurriedly, and with much resentment on the part of the host, chorizo is a delicate lover, spicily sensual, with a warm, passionate softness.) A computer couldn't tell you all that, could it?

Language

R u worid bout loss lang? Dat we wnt no hw2 tlk prply, or wrt gd? Cos tec = tkn ova?

This is the dystopian nightmare of those who claim that technology is to blame for "rising levels of illiteracy" (according to the Literacy Trust, while one percent of the UK population is "completely illiterate", around 16% are classed as "functionally illiterate" - www.literacytrust.org.uk), and that young people think it's acceptable to write job applications in text-speak.

Personally, while my text messages are virtually illegible to anyone who doesn't know me well, of whatever age, unless I make a deliberate effort to text comprehensibly, for example if I'm texting someone I don't know all that well, it would never occur to me to use anything other than grammatical English for letters, emails, job applications, and, of course, writing books. The human brain is fluid, and more than capable of holding an awareness of two – or many – standards of acceptability at once, and switching between them.

When radio first became popular, people didn't lose the ability to read books, nor the desire to attend live performances. Radio didn't fade into obsolescence when people were able to purchase first cassettes, then, later, CDs. It has even survived in the digital download era.

The Bible didn't stop selling, or suddenly become "less relevant" because it was translated into English, or because versions were produced that used a very contemporary lexicon (The Message, for example.)

And just what is 'illiteracy', anyway, given that the language we currently call 'English', the language some people seem so terrified of losing to 'text-speak' and the 'digitalisation of everything', would be

virtually unrecognisable to a 16th-Century
English speaker. An individual watching the
debut performance of any one of
Shakespeare's plays would find themselves
confronted with a host of unfamiliar words –
unfamiliar because Shakespeare had simply
made them up.
 Language isn't a fixed and enduring
concept. It is fluid, constantly changing
and evolving, always bubbling away just
beneath the surface of things.
   Language, and its associated skills, can
never be 'lost', because language is a
function, not a form – it is there purely to
enable human beings to communicate with one
another – primarily, of course, with others
in their particular 'tribe' – and, as long
as that purpose is being adequately
performed, as long as people can understand
most of the conversations and communications
of most of the people around them most of
the time, language remains, however
unrecognisable and incomprehensible some
people may find it.

Community

Another thing people seem to fear losing to
the Internet, to the vast, eternal gods of
technology we have created, is a sense of
community, a feeling of being genuinely
connected to other human beings, whom we are
able to call upon for help, share memories
and experiences with, and offer advice to.
We worry that 'globalisation' and the rise
of the 'digital age' will lead to a loss of
our sense of belonging, a loss of belief in

the idea that human beings are more than just individuals. We are concerned about the 'lives of isolation' many of our young people appear to be living, at the opportunities for 'genuine' interaction we believe them to be missing out on. We worry that, as we gradually come to truly 'know' fewer and fewer people, as Facebook friends replace real-world friends, we will, inevitably, come to a point where we refuse to co-operate with people, because we don't have an intimate sense of connection to them, and, so the argument goes, are not concerned about them.

 In fact, anecdotally at least, the opposite seems to be the case – technology seems to bring about, as well as enabling, a generosity of spirit towards people we may barely know that is sadly lacking in 'real-world' interactions.

 Recently, an elderly man who regularly performs puppetry dances to music had his stereo trashed. Within hours of this being posted on social media by passers-by, he had been given three new stereos and £200 in cash. Had this man simply asked someone for the money to replace his stereo? He would have probably been ignored.

 At 3am, not many people are likely to respond favourably, if at all, to a phone call from someone who is lonely and 'wanting to chat to someone' – online, however, someone is always up and about and ready for conversation, whatever the hour or the subject.

 Perhaps it is because technology doesn't engage us on the same emotional level as physical interaction that we are more willing to respond to a Facebook status than a face-to-face request, that we put more faith in what a person posts on Twitter than what they tell us in a job interview. When you engage emotionally with someone, it can be exhausting – and that exhaustion can lead to an automatic decision not to help, especially if the person you're talking to has caught you at a 'bad time.' Online, you can read people's messages and posts when it suits you, rather than when they happen to barge into your life, and are thus able to make a conscious decision to engage only when you are in a positive frame of mind, and not overloaded with other obligations, anxieties or reponsibilities.

Digital communication, with its lack of emotional involvement, makes it very easy to ask the 'deep and meaningful' questions early on in a relationship – to get to know someone better in six weeks than you might in six months of exclusively real-world interaction.

Communities flourish when there is something to draw people together, and there will always be those who, not understanding the draw of a thing, will fear it, and fear for the community. I don't doubt that the elders of primitive human tribes, seeing fire for the first time, muttered grimly about how it would destroy the community, as people would no longer need to sleep together for warmth, and so could live on their own, away from other tribe members, and yet fire became campfires, with all their songs and stories, laughter, and community.

Security

There are eyes everywhere, and walls have ears.

We've never willingly shared so much information with so many relative and total strangers. Never have so many faceless organisations legally known so much about us, our families, our hobbies or our shopping habits. You don't have to be paranoid, or have over-read *1984* – Big Brother **is** watching, 'they' **do** know where you live (and shop, and go on your lunchbreak), and, in some cases, they really **are** out to get you – although a Trojan computer virus is unlikely to be quite as lethal as its namesake.

The similarities between the technological and historical Trojan Horse, however, are eerily similar.

A couple of bleary-eyed soldiers peer out of the city gates, having been at war for years by this point, and see a random wooden horse, that could only, realistically, have been left there by their enemies, open the gates, and wheel it on through.

A bleary-eyed desk squaddie sees an email from someone they don't know, or an ad for a site they've never heard of, and clicks...

Lazily, we 'click OK' when a website asks if we'd like 'it' (ie, the people behind the programming code) to remember our password. We give out our debit card information and bank details without a second thought, for goods we haven't actually seen, much less received, from a seller we've never heard of. So-called 'trust agents' buy our trust for pence, and betray it for fortunes. And who knows who has access to the records linked to the serial numbers that appear on every UK ballot paper (and next to our name on the Electoral Roll), despite the fact that our vote is held to be "between us and the ballot box?"

Relax - for all its seemingly endless opportunities to shatter our security, technology isn't all bad.

CCTV, and the ease of sharing images online, now makes it easier than ever to track down and identify the lawless and the missing, we are increasingly able to store a variety of documents 'in the cloud', thus removing them from the risks of fire, flood, marauding pets and children, or house moves.

Part of the security of technology is the fact that many people use it, trust in it, and therefore have an interest in keeping it functional. A lot of people with something to lose will result in greater attention and concern, and swifter action to repair damage.

Perhaps one day we will no longer say that something is "as safe as houses" - which have been proved, time and time again, to be incredibly vulnerable to a myriad of external forces - but, instead, say that it is "as safe as servers" - more accurate, even if, at the moment, somewhat less believable.

LEVITICUS

The ultimate proof that all of our current
technological achievements are just another
point on the path of human evolution, and
the evolution of human communication, can
most clearly be seen in the "laws" we
construct, informally and on an ad-hoc
basis, to govern online behaviour, and
interaction with and through technology,
which, for the most part, are incursions of
real-world boundaries, limitations and
expectations into the blogosphere, onto
social media platforms, and, eventually,
into our minds and instinctive, collective
understanding.
  Thus, "it's not polite to shout", becomes
"Please do not use ALL CAPS in posts." In
both cases, there is a practical and valid
reason for the "law" - those who are
hearing-impaired and reliant on lip-reading
struggle when the person they are listening
to is shouting, and, for visually impaired
individuals, screen reader technology will –
literally – shout any word that appears in
block capitals – not ideal in a busy office,
for example. "Just ignore bullies – they
only want a reaction", becomes "Don't feed
the troll" - a reference to the mythical
creatures who, just like their internet
namesakes, are best-known for lurking in the
dark, eager to disrupt the "journeys" of
those above them. Just as we were once
instructed to share our sweets and toys,
and, later, to write the explanation of the
Maths problem we'd solved on the board, for
the benefit of the class, we're now asked to

leave our coding visible, so others can improve on it, to make our content freely shareable, and to link to sources, so others can follow our line of thought, perhaps reaching their own, divergent, conclusions.

Of course, where there are laws there will, inevitably, be lawbreakers – and the realm of technology is no different, apart, possibly, from the increased profile of lawbreakers who break laws not for their own gain, but for the improvement of everyone's experience. Often, these "lawbreakers" seem far more agreeable than the legal entities they are challenging – very few people support Facebook's "Real Names" policy, and, in fact, Facebook have no legal grounds for demanding that people provide ID to back up the name on their account – general conversations and interactions, such as those conducted on Facebook and other social media sites, are not legally binding, and, therefore, there is no legal requirement for formal identification to be produced – in a casual interaction, you can go by whatever name you wish. Only when you are engaging in legally binding activity are you required to use your "legal" name – that's why it is referred to as a "legal name" in the first place, where individuals, for whatever reason, use aliases. Many people who have been unfairly hit by Facebook's privacy infringement, in the form of demands for documents that the shadowy figures now behind Facebook have no legal right to see, are moving to other social media platforms – just as, in real life, human beings will stop interacting with those who are perceived to be making unreasonable demands of us. As human beings, we have always valued privacy and freedom, and it is unlikely our evolution will ever reach a point where this is not the case.

The increasing, and concerning, rise of privacy invasion in the technological sphere is merely a particularly visible manifestation of the contradiction inherent in human beings, and in our evolution – we become frustrated with a particular way of being, announce that we're going to "do it better" - and, inevitably, bring in the detrimental elements of the way of being we claim we want to leave behind.

For example, it is actually more beneficial for human beings to live alone; disease doesn't spread as rapidly, we are less likely to over-farm or over-hunt an area, as we are better able to estimate our own, individual requirements than those of a large group (it is generally accepted that, once a group of anything numbers beyond four or five, all animals, including humans, will typically stop being able to process the individual members of the group, and just see "a lot of -"), and we are more likely to help one another out, as we are aware that "someone else" may not be aware of a person's need. But, focused as we obsessively are on the fact that we have the ability, and apparently innate desire, to engage in complex communication, we have built crowded towns and cities, have destroyed the very countryside that draws people to our shores, have crammed people in one on top of the other, and created problems that didn't exist for our "primitive" ancestors – and then claimed this as an improvement, as an impressive feat of human ingenuity and ability.

Why? Because, historically, we have always lived in communities. We don't seem to be aware that this was because our lives were physically much more demanding, and that we would frequently engage in activities – hunts and battles – that were better suited to groups rather than individuals. The need for communal living is no longer there, but we bring it in anyway, because we can't quite let go of something we've always had. Our brains haven't kept pace with our evolution, and so still need a comfort blanket of "something familiar" in order to feel settled.

"...I made a discovery today. I found a computer. Wait a second, this is cool. It does what I want it to. If it makes a mistake, it's because I screwed it up, not because it doesn't like me, or feels threatened by me, or thinks I'm a smart ass, or doesn't like teaching, and shouldn't be here..." Lloyd Blenkinship, in 1986, perfectly captured, not deliberately "bad" people, but intelligent, curious people who were held back and frustrated, of creativity held back by the chains of a system that has never been interested in personal development, personal happiness, or individual success. A system that, in fact, is designed and funded to fail.

Computers, always the home of the outcast, have now become mainstream, are now being controlled by the same minds and forces that brought the outcasts into being in the first place, and those same outcasts – the very people who were among the first to realise the full extent of what computers could do, the first to set about making computers do the things these individuals knew were possible for the machines, are being labelled as "dangerous" and "difficult", and – as has been the case since their school days – are being left to their own devices, shunned by the slick-suited latecomers who want to believe the bright and shiny things are theirs exclusively. Bubbling cauldrons of creativity, intelligence and frustration are being stuck in corners and forgotten about – something unfortunate is bound to happen, and, dimly aware of this, the slick suited ones set about creating laws, their familiar tool for controlling their environment, and any rogue elements that might somehow find their way into it.

Let's take a moment, and examine why laws are brought into being in the first place – what the likely original motivation behind them is. Let's start with a set of laws that almost everyone will be familiar with – the Judaeo-Christian Ten Commandments.

1.

Love the Lord your God: less the "religious nonsense" it might, at first, appear. At the time the Ten Commandments were written (on tablets, it's worth noting...)the Jewish race were homeless. They had fled a bad situation, and ended up in what, to many of them, seemed a worse one – they had seen friends and family die and be born in the desert, were tired, and running out of faith in this so-called "Promised Land." Having a commandment that put a common focus – the god of the Jewish faith – at the very centre of thought and life was a quick and simple way of binding a desperate, disparate band of people together, keeping them focused on the distant goal, and uniting what, at the time, were most likely familial tribes given to bitter and potentially violent disputes, as is still the case in much of the Middle East today.

2. Love your neighbour: the original "share your code, and help a fellow techie". The story of the Good Samaritan, where a man from Samaria – a "heathen" country whose inhabitants were scorned by the Jews – is the only person to stop and help a Jewish man who has been beaten and robbed, was told to make clear who your "neighbour" was – in Biblical and technological terms, your "neighbour" is anyone who helps you. It is likely that this commandment came about in order to

3. encourage co-operation in the early stages of building a viable society – something that always experiences a fair amount of teething problems.
4. Remember the Sabbath Day, and keep it holy: "Take a screen break." The age-old, intuitive understanding that people work better, and have their best ideas, when they're allowed and enabled to rest.
5. Honour your father and mother: remember and respect those whose hard work has made your life possible. A commandment that would, in a time and circumstance where resources were limited, and competition fierce, ensure a reasonable level of inter-generational harmony, and – particularly important in a desert climate – cut down on the number of bodies needing to be buried as people were killed because there was "no point" to them any longer.
6. Do not lie – technological worlds, as much as real-world societies, work on trust. In the technological sphere, it is easier than ever to "dox" someone – to find out everything about them, to follow their digital footprint, etc – but, also, it is easier for individuals to manipulate that footprint, to create a highly functional and persuasive illusion. Words, in this world as much as the Biblical, are a currency and a power; encouraging honesty encouraged a society in which interaction, commerce, and harmony was possible – vital for that society's long term viability.

7.
Do not covet anything that belongs to your neighbour: omit the apparently confusing references to oxen and asses, and you're left with a guideline aimed at producing people who are content with what they have, even if that isn't much. You have a society that will cohesively stand together, rather than risk being torn apart by material desire, and the jealousies it creates. A good idea, whether you're religious or not.

8. Do not commit adultery: in a society that is heavily invested in inheritance, as Biblical societies were, this is a vital edict, as men needed to be certain the children they were passing their estates to were, in fact, theirs. It is likely that this also limited the chance of someone inadvertantly having children with someone to whom they were too closely related, as marriages would have been arranged in order to avoid too-close relations, as far as possible. In the technological world, where content is created for kudos rather than cash, and brands are built on personalities, this commandment is still important. People need, and have a right, to know whose "baby" something is.

9. Do not cheat: another commandment related to creating a successful society when the major currency is a person's word and their reputation – as true of the 21st century digital realm as it was of 1st century Palestine.

10. Do not kill: at first glance, you
    might wonder how that links in to
    the technological world, where
    physical presences that can be
    killed are markedly absent. True,
    you may not be able to actually kill
    a person (well, outside of RPGs,
    perhaps),but you CAN kill their
    hopes and dreams, their creativity,
    their ambition, and their courage –
    and, when you're trying to build a
    better society, ensuring as few of
    those things that "die" as possible
    is as important as it is to keep as
    many individual people alive as it
    is when you're building a society to
    begin with.

   "The letter kills, but the spirit gives
life" - this is the apostle Paul, in his
second letter to the embryonic church in
Corinth. (2 Corinthians, 3:6). As Paul
correctly identifies, laws that are made
thoughtfully, with a "good spirit"
encourage, enable, and protect people – they
give life. It is the laws that are made with
an agenda in mind, laws that are made for
the sake of making laws, laws that are made
with the sole intention of punishing
particular groups of people, or particular
individuals, that "kill" people – turning
them into soulless automatons, uninspired
and uninspiring.
   When you realise that the virtual world
offers all the limitless possibilities of
the spirit of laws designed without agendas,

you realise just how much of a misconception
it is that "the internet is a magnet for
anarchists and terrorists." It isn't, and
can never be. Anarchists don't want laws –
they don't want society to prosper, they
want it to fall apart. They don't want
stability – they thrive in the flux of
chaos. And terrorists want the stifling,
murderous letter of their own laws. With its
bright, brilliant spirit of law-based
freedom, the internet will never be a home
to either anarchists or terrorists, but only
ever a resource; never a magnet, but always
a "necessary evil."

It is worth noting, as a final aside, that
Moses, in a fit of pique, is reported to
have thrown the stone tablets on which the
Ten Commandments were originally written to
the ground, breaking them – the Biblical
equivalent of hurling our contemporary
tablets at walls, in frustration at the
stupidity of our colleagues, or various
individuals we encounter on social media,
perhaps?

NUMBERS - A CENSUS OF THE INTERNET.

Possibly the most famous statistic about the Internet is the revelation that, if Facebook were a country, it would be the third-largest in the world.

That sounds pretty impressive, until you take into account the fact that this is a 'country' where the cost of living is merely the ability to connect to the Internet, even if that is only via public computers in schools, libraries, etc, a 'country' with no citizenship or residency tests, and, finally, a 'country' that has almost completely open borders (bearing in mind censorship issues and restrictions in certain nations)to and with over 100 other countries (there are 196 recognised countries in the world, it is likely that not all of these have Internet access, however.)

Does it still seem impressive, that statistic?

The point of this example is not to mock the undoubtedly massive membership of Facebook, or any of the other impressive numbers and percentages that will come up in relation to technology, and particularly the Internet, but to remind ourselves that it is very easy to get a "bubble-view", where we can't see past the *fact* of technology to the setting and conditions behind and around it. We can get caught up in thinking that technology IS the world, rather than remembering that technology is merely *part of* a far wider and more complex whole that *is* the world.

So, who makes up this country of the

Internet? Well, North America accounts for over 84% of internet use (84.4%, according to internetsociety.org as of 2013), while sub-Saharan Africa has a small, but, given that area's situation, impressive, 17% of Internet users. There are over a trillion Web pages, and Kevin Duncan, in his book Revolution, explains that the Internet is only currently capable of storing HALF the content that has been, and is being, uploaded to it – the rest is "in transit" somewhere in the ether.

What are we all talking about, on those one trillion Web pages? Of course, the first response is likely to be "porn", since almost everyone believes that "the majority of the Internet is pornography." (It probably is, although the forms of pornography may be much more subtle than your straightforward sexy/kinky sites – so-called "inspiration porn" is currently big business, and it is likely that there are many more sites dedicated to "how to improve your life!" than there are "drool over these sex objects" sites. For example, the BBC recently suggested that the percentage of porn on the Internet, when one analysed *sites*, rather than pages – of which pornography sites would have rather a lot, because of their dependence on giving viewers something new each time – pornography accounted for just 4% of the Internet – which, as Jamie Bartlett points out in his book DarkNet, is, in relative terms, about the same percentage as pornographic magazines have in that sector.

152 million websites are personal blogs, according to Wikipedia (2013) – but, as Wiki notes, a high percentage of these sites, whilst still "active" in the web's eyes, are, in fact, abandoned by their owners after the first few weeks or months – as we either get bored of talking about ourselves, to a potentially mythical audience, or invent new selves to talk about. As a recent contributor to Huffington Post's Young Voices site admitted, "social media is a carefully constructed farce..." Those of us who dare to put our real, unedited lives on Facebook, for example, will often very quickly find ourselves just watching the tumbleweed roll by – people don't want reality; they WANT the airbrushed fantasy, however much they may claim otherwise. Real life isn't inspiring, unless you're in a position of being able to simultaeneously pity the person whose life is being portrayed, whilst feeling smug about how much better your own life is. This works best when you have little or no connection with the recipient of your pity – when someone you see down the pub, or work with, or live near, starts talking about their health conditions, their unemployment, the caring responsibilities they have for a parent, a spouse, or a child, it becomes intrusive, because that voice in your head starts whispering that you should maybe "do something to help them out" - and, because the human brain is designed to avoid change, that's an uncomfortable thought.

It is an almost-daily gripe, in various local and regional newspapers, that rural areas don't have the same quality of access to superfast, fibre-optic broadband as urban areas, but, again, this needs to be put into perspective – it took 35 years – three and a half decades – for everyone to be connected to the landline telephone system. Broadband achieved full connection (although not always to high-speed lines) in just SIX years. But, of course, the rise of technology has seen a corresponding decrease in our patience – as Simon Cowell pithily observes, where we once used to be perfectly happy to queue up for half an hour in the bank to take out £50, now, if we're standing behind someone at an ATM who's printing a receipt, we "...actually want to kill them."

I've worked in offices – I've seen (and heard) the fury of colleagues if they can't get an answer *right now!* I've also seen the meltdowns that occur with the realisation that "the computers are down." (Meanwhile, I'm annoying the hell out of everyone by going to my hard-copy diary, looking through my hard-copy files, and getting on with whatever I can manage without a computer – I actually got sacked from one job for keeping these hard-copy backups, as "everything has to be instantly accessible to everyone else in the office – that means diaries, records, etc should ONLY be on the network!" Pointing out that, by keeping hard copies, I was able to get on with my work while everyone else was sitting around waiting for IT to show up, didn't help.) In losing our patience – in taking the human out of so much of our day-to-day lives – we've put our humanity at risk. It is, undoubtedly, useful to be able to find out, in under a minute, how many countries there are in the world, or what percentage of the web is actually pornography, but the world wouldn't end if I had to find those, or similar, stats some other way – this book might take longer to get written, but, on the other hand, maybe not – if I had to look things up the old-fashioned way (I refuse to call it "the hard way" - it really isn't that hard), I probably wouldn't be distracted by avenues that were nothing really to do with my original question, I certainly wouldn't be tempted to wander over to a random group of people in the corner of the library and start chatting to them, and I wouldn't have to take a break every couple of hours because my head was aching and my vision was starting to blur. (Although I'd run across a bit of a problem when it came to getting this book "out there" - my handwriting looks like a spider got drunk and crawled through some ink, before staggering across the page, and I have a roughly thirty-minute window before even I start to be unable to read my own writing.)

When we look at the numbers of the digital world, we start to see future points on our evolutionary path. For example, in the UK, there are 25.419.296 TV licenses registered, for a technology that dates back to the late 1930s, and which took 26years to reach the level of popularity it now enjoys. Netflix, which was set up less than twenty years ago (1997), currently has 3million subscribers. It is impossible to tell how many people (like myself) don't have a television, or a

Netflix subscription, but do watch television online, either via catchup services from terrestrial broadcasters, or via YouTube. (I love the fact that The Dresden Files, and old episodes of The Bill, are easily found and watched via YouTube.) It is likely to be significantly higher than the 3million registered Netflix users, and, as the debate about the BBC's UK license fee rumbles on, is probably only going to increase, as more and more people decide to stop paying for what, at a domestic level, is often seen as "constant repeats." Perhaps, in another twenty years time, television, in the form of a box with an aerial that you have to pay a license fee for won't exist, in the same way the analogue signal for television doesn't exist, and "big back" TVs are already disappearing.

This paradigm shift is a double edged sword. On the one hand, it opens up a meritocracy of success for creative types – as the Internet replaces traditional media (as it almost inevitably will, barring some catastrophic incident that either crashes the Internet, or radically alters peoples' thoughts on technology), people will be enabled to showcase their work and their talents to the world without the barriers of auditions and connections – in time, the teething problems around a lack of quality content that are currently being experienced, particularly in the self-publishing sphere, will be ironed out, as viewers and readers adapt, and learn to engage with those who have genuine talent and ability, rather than simply "anyone who's out there" - the poor-quality videos, tracks, and books will disappear, as their creators stop bothering once their views and/or income tails off.

On the other hand, it brings ever-closer the day when housebuilders and governments decide that people "only need" a tiny, pod-like space – a situation that is already fact in Japan, and which has been discussed previously in the UK, with one architect, who is adapting traditional shepherds' huts for modern living, observing that "...younger people don't have 'stuff' any more – their books are on a Kindle, their music's on an Ipod – pretty much all they'll really need in a few years is a bed, a bathroom, a kettle and microwave, and somewhere to plug a charger in."

People – particularly people focused on maximising their profits – forget that human beings are not, in fact, fully social animals; like cats, homo sapiens is a species comprised of "individualists who co-operate." As housing estates like Tower Hamlets prove, we don't do well when we're too close together, with too little space to call our own.

And what happens to the people who can't afford to "keep up with the Joneses" on the technology front, or who have other interests beyond music and technology? I don't have an e-reader – I have books. I also have pets, a feature that is fairly central in my life – four large breed dogs certainly wouldn't fit comfortably in a shepherd's hut, however well converted and

adapted the design was! Other humans enjoy
activities such as scuba diving or canoeing,
for example, which involve a lot of bulky
equipment that needs safe storage. Will
those people end up forced to chose between
necessary shelter, and the hobbies that
promote physical and mental health,
encourage genuine, real-world interaction,
and provide a quality of life? If they are,
expect the costs to the NHS (assuming it
still exists) to spiral, as levels of
obesity, stress, depression, and all the
illnesses and complaints associated with
them rise, as people are cut off from the
health-giving benefits of interests that
aren't based in a screen world.
 Kevin Duncan, in Revolution, observes that
"...Indians are more likely to have access
to a mobile phone that a lavatory...",
proof, if we needed it, that society,
insanely, is moving towards a point of
valuing technology and gadgets – which can
only ever be "nice-to-haves" over the
necessities – such as functional sanitation.
Technology, and its massive rate of
progress, are causing us to lose all sense
of proportion, and all understanding of what
our needs really are.
 In fact, many people believe that
technology has infantilised us – we're so
used to some faceless tech corporation
telling us, via clever marketing, what we
need now, that we're losing the ability to
think for ourselves. They point to
teenagers, engaged in their silent worlds of
virtual communication, "hanging out" in
their rooms with "friends" they've never
met.

At the age of 14, Joan of Arc was leading the French Army – what is today's typical 14 year old doing? Probably not even leading a "Twitter Army" in a passionate social justice campaign. And what does this mean for the future of our workforce? British business, in particular, is already complaining that even graduate-level candidates lack basic professional social skills, maturity, and the ability to use their own initiative to solve a task. The British government has already begun subtly moving back the age at which individuals are considered "functional adults" – education is now mandatory until the age of 18, housing benefit can't be claimed, other than at "shared room rate" (which, in many areas, is less than a "shared room" actually costs), all of which means the age at which an individual is typically able to buy their own home (and thus fully commit to a course of employment or education, or, indeed, get married) is increasing. It also means that the people are typically leaving it longer to start having fewer children, which is storing up a potential time-bomb in terms of the availability of a workforce able and willing to care for an elderly population in which increased life expectancy – due in no small part to advances in technology – has given rise to a higher incidence of Alzheimer's and dementia, along with more, and more serious, physical frailties.

Bear in mind that, at the current point, 99& of Internet users come from white, middle-class homes, with the majority of those users living in urban areas (internetsociety.org). These people, having current access to technology, will be the first to have access to *future* technology, the first to understand the new paradigms that are coming through, and the first to use those paradigms to their advantage. Inevitably, these people – these "digital natives" - will shape the form and content of, and access to, future technology, along with the way that technology impacts upon future employment. The Internet may give the illusion of offering greater access and impact to minorities, but, until Internet use among minorities – including economic minorities – reaches parity with that of the white, urban-dwelling middle-classes, it will only ever be an illusion.

Technology does offer genuine value, in many cases opening up avenues of employment to people for whom physical location, lack of transport, illness or disability would have previously been a very real, and often insurmountable, barrier, and, of course, enabling the kind of interaction and near-instantaneous connection seen during the Arab Spring. It enables the forces of authority, and of law and order, as well as commercial enterprise, to be held accountable for unacceptable actions and attitudes, and it enables more people to contribute to humanitarian efforts, national and local causes, and individual pleas. However, it is easy, particularly for digital natives, to lose sight of the necessity of pre-existing social capital required for technology to be *successfully* used in this way. As every event organiser of the digital age knows, often to their cost, the number of people who have listed themselves as "going" to an event on a social media page rarely translates to the number of people who will actually turn up. And if you don't have the "capital" of a good network already, it can be very hard to get *anyone* to join an event, sign up to a newsletter, or buy something you've created.

DEUTERONOMY

"Deuteronomy is organised as a series of addresses given by Moses to the people of Israel...when they were about to enter and occupy Canaan."

We are almost 40 years into the existence of the Internet (the basics of what would be first the ARPANet, later the Internet, were laid down in 1978), and almost 80 years into life with computers – the first computer as we would understand them was invented in 1938. Just like Moses and the Israelites, we have spent our time in the wilderness, and are about to enter and occupy our Canaan, the land of "milk and honey" of easy-access data, instant connection and gratification, and the land where, while we may still not achieve instant fame, we can instantly "put ourselves out there", and increase the likelihood that fame, or its representatives on Earth, will see us, like us, and come knocking.

The first thing we need to look at is where we are – what is the nature of the "land" that we are on the verge of "occupying"? What "milk and honey" does technology offer, and what threats do we have to neutralise first?

Technology offers us unparalleled opportunities for progress, a once-in-a-lifetime opportunity to completely change our concept of what work is, and how it is carried out. People in distant countries, who previously would have had no power over oppressive regimes there, can now get the attention – and the assistance – of the most powerful nations on Earth. Individuals without resources, who would have previously had to struggle on, to "make do and get by" are now able to reach out to hundreds, thousands, sometimes millions of people who are all able to do one small thing – which, taken together, add up to the big things. It is easier for people to find new job opportunities, or to create their own business, product, or market place.

   And yet, as the world becomes increasingly reliant on technology, it becomes easier than ever for despots and tyrants to shut their countries off – simply by banning technology, or making access to it all but impossible for the average person, they bring up walls that, unlike their physical counterparts, cannot easily be bombarded and brought down, or clambered over, walls that will never bear the marks of spray-painted rebellion, walls that will be forever silent in the face of the oppression they shield. As the rise of the Internet brings about the "cult of personality", and an increasing reliance on "social capital", people without the ability to "self-market", without a story, and the ability to tell it, people who aren't photogenic and don't have the kind of experiences and views that are rewarded with vast numbers of Facebook friends and Twitter followers, will be left to struggle on, increasingly desperate and despairing. As the numbers of Change.org petitions and crowdfunding appeals rises, genuine causes, and people in genuine need, will get pushed under by the rising tide of "nice to have, but could live without" causes and campaigns. As workplaces become increasingly digitalised, not only are those unable to afford the latest technology pushed further and further back in the queue of success, those *in* work experience higher levels of anxiety and depression, are subject to increased demands on their time, energy, and attention, because technology, in removing the *negative* boundaries, of secrecy, opaque processes, and "appropriate complaint channels", has also removed the positive, essential boundaries, such as the understanding that it isn't acceptable, unless it is a genuine emergency, to call someone on a Sunday, or after 9pm, or the boundary of respect that says bad news should be delivered face to face, that interpersonal issues should be dealt with between people, not between their email accounts. The boundaries of "appropriate complaint channels", while often used as a shield by guilty parties, did serve a genuine and positive function – they prevented management from becoming so overwhelmed with (often quite trivial) complaints and grievances that they had no time to deal with anything else – including the day-to-day running of their company. For this reason, officially at least, FirstGroup, which runs bus services in the UK, has removed itself from social media interaction. Perhaps more companies – and perhaps particularly the poorly-performing companies – need to do the same. In a world that is "always on", perhaps we need to switch off for a bit.

Technology has granted us unprecedented amounts of "leisure time", and yet we don't really understand what "leisure" genuinely is. As Ricardo Semler, a Brazilian businessman, points out, leisure is NOT the opposite of work – the opposite of work is *idleness*. And, not understanding this, we allow all of our technology to use us. We allow it to facilitate "working" during our leisure time, because we believe that leisure is the same as idleness, when, in fact, they are distinctly different entities. Why can we answer emails on a Sunday, but we not go to the cinema Monday afternoon? Because we believe that answering emails is always "productive", and we fail to see that going to the cinema can sometimes be *more* productive than sitting in the office staring at paperwork, or slinging yet another box of components off the factory line. In evolutionary terms, humans are a young species, and the concept of "leisure time" has, relatively speaking, only been here five minutes. Our brains haven't caught up to where technology has taken us, and, not understanding, we feel guilty. We are programmed to "work" - to undertake action that leads to the fulfilment of a necessary requirement, or to a positive and pleasing outcome that improves our status or situation – and so we feel awkward about having "all this leisure time" – because we don't see that leisure IS work; it is, at least when used well, the kind of work that leads to a positive and pleasing outcome that improves our status or situation.

In my own leisure time, I read, spend time outdoors with my dogs, take photographs, and walk by the sea. I'm working at exposing myself to new ideas, new ways of thinking about old ideas, and new worlds, at seeing things differently, at learning how to interact with entities that don't communicate in the same way I do, and I'm working at seeing things differently. These are all positive and pleasing outcomes that, in small ways at least, will improve my status and situation, because they all develop the "soft skills" that are necessary in the world of paid work, but that, unlike MS Access, Six Sigma, or the latest manufacturing process, can't be taught.

 Time and again, studies have shown that, when we're faced with too much choice and not enough reliable information, our brain – still being a relatively young entity – helpfully hands decision making over to the limbic system – the reptilian hind brain, the ancient part that functions on emotional reactions, primarily to feelings such as fear and anxiety. In the modern world, where – in the West at least – we are unlikely to be killed by competitors or predators, the limbic system rarely makes *good* decisions.
 You would think that an increased amount of technology, and increased access to that technology, would mean there was more reliable information around, whether in the form of news broadcasts, newspaper reports, online "citizen journalism", or simply easier access to records and data, meaning

that the brain would *stop* assigning important tasks to outdated and incompetent areas, but, in fact, the opposite seems to be the case – more and more people are making more and more decisions based on irrational assessments and reactions – because, although we have a veritable flood of information, it is incomplete, because we don't know the agendas of those behind it, we often can't track down where it originated from, it's more susceptible than ever before to misinterpretation and manipulation, and, particularly in the case of the Internet, we're less able to identify what is *actually* new and useful information. Technology is a point on the path of human evolution, but, in many ways, it seems to be a point we've reached by taking a short-cut. Had we taken the longer route, we may well have arrived at technology with brains that had evolved enough to handle "information overload", and all the other apparent perils of the digital age. As it is, we missed that bit, and will have to operate as best we can with that cognitive lag, while we wait for our physical evolution to catch up to the technological one.

"Technology is dangerous, because it exposes children to violence."
Children have always been aware of violence – who can't readily picture little boys, either themselves, their brothers, sons, cousins or male friends from childhood, playing cops and robbers, cowboys and Indians, soldiers, etc? Guns have always been popular toys, playfights have

always been one of the many ways in which young boys exercised growing muscles. There is an instinctive draw in all humans towards violence – Thanatos, the "death drive", which co-exists in a kind of tense harmony and power-play with Eros, the drive to live, love, and continue our genetic existence. If we are never allowed to explore Thanatos through play, it will become a part of our shadow self, finding more sinister outlets in our adult lives, and possibly causing serious and lasting harm, far beyond a few bumps, bruises and stitches, along the way. The problem isn't that technology exposes children to violence – the problem is that, as a society, we assume that technology can "provide all the answers", and respond with panic and censure when we realise that it can't.

Let your children see violence on screen. Let them read about it in the newspapers. Let them act it out in physical play, or through video games – but be there to explain the ramifications of violence to them. Be there to teach them about the history of violence, the lasting effects that war has on combatants, and the civilians who live through conflict and occupation. Be there to discuss alternative ways of achieving and maintaining power. Be there to explain the difference between leading through fear and leading through respect.

When your children are young, you should be a god to them – do not allow your place to be taken by technology.

"This invention will produce forgetfulness in the minds of those who use it, because they will not practice their memory...you have invented an elixir, not of memory, but of reminding, and you offer your pupils the appearance of wisdom, not true wisdom." This was Socrates, talking about the terrifying, society-shattering invention of writing. Looking back from a point of evolution from which it is impossible to imagine a life without writing, we can see that Socrates' fears were mostly unfounded – we still value genuine memory, even though writing is an everyday thing that we do without thinking. This valuing of memory can be seen in the fact that we worry when we "start to forget things", fearing Alzheimer's. It can be seen in the furious anger of a boss or parent whose employee or child has forgotten an important meeting, or a certain necessary item. And we can see it in the shame that employee or child feels at the forgetting.

The invention of writing as a means of communication was a paradigm shift, in the same way as the invention of technology was, and continues to be. In every paradigm shift, something is lost. With the advent of writing, it was what Socrates called the "dance of dialogue", that art form of discourse where, with the rapiers of wit and knowledge, holes were poked in theories, and sacred cows of opinion or belief were slain. Perhaps the invention of writing was truly the invention of lying, the beginning of people simply taking words at their face value, as they lost the tone, facial expression, body language and emphasis that

would, previously, have informed,
influenced, and flavoured dialogue.

What we are seeing, at this point in the
evolution of human communication, is the
realisation that the written word (which is
still the driving force behind all
interactions with and through technology)
facilitates lying, and, therefore, an
increasing awareness of the importance of
learning to go behind the words, between the
lines, to find and engage with the truth.
Just as people now write without thinking,
so, one day, they will, without thinking,
identify the authors and agendas behind
words, before they make decisions or take
actions based upon those words.

How can we promote and ensure equality, as
we follow this path of evolution, hand in
hand with technology? How can we ensure that
these new gods we've created don't just
favour those of us in whose image they were
made?

The most important thing is that
technology, and access to it, needs to be
made as easy, and as free, as religion, and
access to it. Just as anyone can walk into a
church, a mosque, a synagogue, or a temple,
if they follow the gods (or godesses) that
those institutions represent, just as anyone
can get hold of a copy of the important
texts for their religion, at an affordable
and reasonable cost, or, if their
circumstances are severely straitened, free
of charge from a leader of their religion,
so it needs to be with technology, with the
new gods. Religion, when in the right hands,

is not pursued or offered for profit; neither should technology be.

The Internet is rapidly taking over – perhaps has already taken over – from religious groups in providing education and furthering social justice – YouTube has TED talks, there are free online courses in any subject you can name, we can quickly and easily see conflicting views, hear dissenting voices, on any given issue, people are increasingly turning to the Internet as their first port of call when they have a question, a problem, or a crisis.

Technology is not just facilitating business – technology is creating business, technology *is* business. The paradigm shift has happened, and the consumable has become the culture. And culture has always been free, has always been open-source and open-access. It's time for us to catch up to that realisation, and act on it.

TECHNOLOGY'S PEOPLE

It is generally held that, without the work of Nikola Tesla, including the Tesla Coil, and Tesla's arguments with his mentor, Thomas Edison, which led to the creation of alternating current as a method of transmitting electricity over large distances, none of the modern technology we take for granted could exist. Of course, if you take that premise, then, in all probability, Tesla wouldn't have had the success he did with his ideas and inventions had he not been sent to Edison, and had Edison not been so convinced that direct current was the only feasible way of transmitting electricity, so we may well, in fact, be looking to thank Thomas Edison, or, at least, to thank him jointly with Tesla – something the latter would have hated; in 1915,Tesla refused a Nobel Prize because it was being awarded jointly to him and Edison.

Nikola Tesla's work paved the way for almost all contemporary electrical gizmos and gadgets, and it was on Tesla's base of quick and simple electrical current transmission, and, indeed, on Edison's base of electricity being something that could be created and harnessed at all, that Alan Turing was able to build the machines that would come to symbolise him, and that would be the next point on the evolutionary path of human beings and human communication. Although now most famous for his work on Enigma during World War Two, Alan Turing was much more than a codebreaker. A King's College, Cambridge graduate, he was also a philosopher and mathematician, and one of the first of the new breed of computer scientists.

From computers and Alan Turing came Vint Cerf, Robert Kahn, and the Internet, first in the form of the ARPANet, with its boards, forums, hackers and trolling, and, later, through the intervention of Tim Berners-Lee and his development of the World Wide Web, to the Internet we know now it. A quick glance at Berners-Lee's page on the website www.w3.org, the home site for his company, shows, interestingly enough, a man who seems almost afraid of the technology he works with, a man who, despite creating an entity which connects literally millions of people, appears desperate to avoid actual interaction with people. This is often the case; the people behind technologies that facilitate mass communication come to fear the near-constant demands that their technology enables other people to make of them; it is almost as though, on some level, these modern day Dr. Frankensteins know that they've created a monster.

Meanwhile, in a recent interview published in the Guardian newspaper, and also on the BBC's website, Vint Cerf is also discussing fear – the fear of a possible "digital Dark Age", where, because of obsolescence, the speed of technological progress, and the difficulties inherent in "backwards compatibility", all of the data currently on hard drives or in the cloud may be rendered unintelligible to future generations. And, given that many people no longer keep any "hard records" – pen and paper journals, printed and bound photo albums – we, the digital natives and the successful interlopers, risk becoming part of what Cerf refers to as "the lost century"; if there are still people on Earth a thousand years from now, and if those people are still interested in the lives of the people who went before them, they may have no way of viewing any of the documents of our lives. In an effort to combat this, Cerf is currently working on so-called "digital vellum", a process similar in its theory to the way museums protect their ancient artifacts.

Perhaps, despite all of the outrage about "semi-literate toilet paper", this is the point of the rise, facilitated, of course, by the Internet, of self-publishing; just as history values the poorly written documents of the lower classes as much as the epistles of royalty and governments, for its ability to show the general standard of education, wealth, and knowledge, as well as what life was like for the average person – vital for sociologists, and those tracking socioeconomic trends, and the broader impacts of government policies. Future generations, if they are interested, should be able to study the lives of the quotidian worshippers of gods, not just their privileged, erudite creators.

Perhaps, though, future generations won't be interested in us in the same way we are interested in the people that went before us. As Peter Drucker, self-styled "social ecologist", as well as a professor, writer, and management consultant, suggested that the Internet "...will bring about the demise of universities as we currently understand them." With a rise in podcasts, TED talks, blogs, and free-to-access courses and resources, perhaps future generations will study history, not through rigid, formulaic working, and not with a specific goal in mind, but simply in an ad-hoc, personal way, studying those things that interest them, working with what's readily available, rather than going looking for something else.

If this is to be the future of study, the basis of how future generations will understand us, then it is vitally important that those in minority groups aren't simply left to "have their voice" through the medium of the Internet – because, if we DO enter a "digital Dark Age", their voices – with their only expression being the online platforms of blogs and social media sites, risks being lost forever. We need to call a halt to the near-obsessive focus and reliance on technology that is currently consuming schools and governments, as well as individuals, lest the god die, and take all of us with it.

You see, a striking thing emerges when you start looking into the technology's people; while not all of the pioneers were American or British, they were ALL white men. And, while white men may have developed the technology we have now, you can be certain that they didn't rely solely on an inherently ephemeral and passing phase of communication to record and convey their thoughts. They will have kept hard records. Their voices will shout out through any digital Dark Age, will echo in the silence that inevitably follows the death of a god. It's the minorities who can only tell their stories through digital platforms, the ones we shoot down because they "don't write proper English", because they don't understand grammar, because their spelling is appalling, because they don't have letters after their name, or a piece of paper brought (usually) with someone else's money, with the name of an "acceptable" university on it, who will fall silent, and, through lack of records accessible to future generations, drop off of the pages of history.

The one thing technology claimed as an absolute victory – an equal platform, equally accessible to the disadvantaged and dispossessed – is the very thing that threatens to disadvantage and dispossess them utterly and forever.

In the realm of technology, uniquely, the worshippers of the god are also creating the god. Where this happens in religion – in fundamentalist versions of any major religion, or in the Christian heresies, for example – other people are quick to point out what's happening, to shout the rogue creators down. Not so when it comes to the god of technology – perhaps because when those worshippers become creators, they tend to improve the religion, not just intrinsically, but also in terms of its interaction with the outside world. The more blogs and YouTube videos that are posted, the more other worshippers realise that they, and their views, can shape the god. But, equally, as the Internet, and social media sites in particular, come to be heavily dominated by those with left-wing views and a commitment to social justice, and as those with dangerous right wing views become increasingly distrustful of technology, increasingly fed up with being targets for the vitriol and trolling of their opponents, the enduring narrative that's being created, away from the temple of the god, is, potentially, a dubious one. I, for one, do not want future generations to believe that the majority of people in the 21st century were violent bigots; I want there to be a balanced view. If there are historians in the future, I want them to see that people tried to combat prejudice through both education and direct action. I want them to see the snarl-ups we had along the way, as people who, for a long time, had been made to feel fearful and vulnerable, still feeling raw in their relatively new skins of visibility, lashed out, with defensive hostility, at those who, while trying to improve things, trying to facilitate dialogue, looked too much like those who had previously stripped away the rights and the dignity of these emergent minorities. I want them to see the hurt that was caused as people were attacked – verbally, rather than physically, and by individuals, rarely by gangs, and never by the forces of law and order that should have protected them – for things they couldn't help: being white, being male, being heterosexual, not being (or, in some cases, not *appearing* to be) transgender/transsexual. I want them to see the rise of "alternative beauty", and the realisation, in the print media world, that unrealistic images of airbrushed "beauty" were no longer desirable. I want them to know that the 21st century was the century of self-determination, where people, not parliaments, dictated whom they would love, and how, what work they would do, and when, what they considered "beautiful", "attractive" or "desirable", and why. I don't want it to seem as though we were silent in the face of injustices committed by governments against their own people because there were no hard petitions, and the records of the Internet actions taken in response to the issues of our day have been lost. I don't want rich white men to carry on telling the story of us, with their voices drowning out everyone else's. And I don't want the story of us to be lost in shouting and snide comments. I want us to sing together, and for our voices to rise above and beyond the god that facilitates their harmony, to become a lasting sound in the world.

   The god of technology should, if it is to exist at all, be an icon, not an idol – something that inspires awe, that people labour over in love, and with devotion, rather than something before which people should just fall down, dumb with wonder.

THE SONGS - DIGITALLY REMASTERED.

As I write this, I'm listening to Depeche Mode, via YouTube. No cost, no waiting for the CD to arrive, no frustration when a bit of dust stops it playing. This morning, before settling down to write, I watched a couple of episodes of The Bill, a police drama series I used to enjoy, and that broadcast its final episode several years ago. This evening, I may well watch The Dresden Files - I love the books, written by Jim Butcher, but, without the Internet (and a broadband connection) would have been unable to see the television adaptation without paying a small fortune for the privilege of a satellite TV subscription. I'm also currently pursing a free course in Art History through OpenLearn, and have a couple of courses coming up, also free, through FutureLearn, in economic theory and computer coding. I am able to quickly and easily find and purchase a book I'm made aware of while I'm reading a different book, and to get some background information on someone who's quoted or referenced in something I'm reading for work. At the click of a mouse, I can see the work of exceptionally talented amateur photographers, all of whom work in my area, find and connect to people with similar thoughts and experiences to my own, and encounter new voices, and new songs. Technology has opened many doors, allowed much more music to be heard by far more people.

And yet, as journalist Sharon Griffiths pointed out in a recent column in the Eastern Daily Press regional newspaper, the personal, intimate art of letter-writing is dying out. It is very rare for people under the age of sixty to have the thrill of reading a letter written in a familiar hand, or the experience of knowing from that handwriting exactly what frame of mind the writer was in as they wrote the letter. I have two peoples' handwriting, besides my own, that I recognise at once – one is my wife's, and the other is that of a good and long-term friend. But I am aware that previous generations would have recognised the handwriting of as many people as I can recognise by their voices. I am aware that I have never received, nor, in all probability, every will receive, a handwritten love letter – it may well prove the case that in my lifetime the practice of sending Valentine's cards disappears – I already know many people who don't bother. I am a writer, and, to me, there is something special about taking the time and the trouble to write by hand. The only reason I don't do so more often is that my own handwriting is virtually illegible, even to those who know me very well – my wife has the same problem; our handwriting is recognisable, but utterly unreadable.

It is certainly a good thing for schools to teach children about technology – to return to the roots of their existence, as places which would instruct the young in the proper worship of their gods – but they should also teach the "old-fashioned" skill of penmanship, too. Because, that way, you have a generation of adults – of future workers – who can function effectively when a power cut or a crash puts the technology they rely on out of action. You have a generation of people who understand the joy and satisfaction that comes from taking care and time over something. You have a generation of people who learn to think before broadcasting their words. And, because those who are taught "old fashioned" skills alongside the skills of the digital era, they understand what life was like before the god of technology, and thus make much more committed worshippers.

Many job applicants of my age (late twenties) and younger express incredulity when a job advert requests that they "submit handwritten applications." Some snort with derision, assuming the employer will be using the "nonsense" of graphology. Perhaps they will. Or, perhaps, the job involves regularly taking notes by hand. Perhaps the employer wants to be assured that they're hiring someone who can actually write coherent English without the need for a spellchecker. Perhaps they're simply looking for someone who isn't put out at being asked to make a bit of an effort, and take a bit more time and care than they usually would. Perhaps they're sick of receiving copied and pasted template cover letters, with a CV that's clearly been written by a Jobcentre advisor, rather than the candidate themselves?

I recently read a friend's pregnancy announcement on Facebook – now, my friend and her partner are both busy people, with a lot going on, and not a lot of money. I appreciate that Facebook is a quick, no-cost way to ensure everyone who would be interested knows the news. But, somehow, it renders mundane something that should be special. The very fact that my friend had to use a clever headline, in block capitals, to ensure people actually read her post, rather than scrolling past it, saddened me. When we're sharing life-changing news with people, good or bad, we shouldn't be having to think of the right "hook", we shouldn't be worried about people scrolling past in search of "something more interesting." It sometimes seems that, in the digital world, we have more opportunities for connection, but less willingness to engage. We're losing the chance to be able to remember, with a smile, the exact moment we heard news such as the expectation of a new arrival, or a marriage date, or someone telling us they loved us. Do we really want to lose the magic of those moments under a torrent of "I was just scrolling through..?"

And then there's the trivialisation of news. At the time of writing, the UK is engaged in mass hysteria (or so it seems) about the rise in popularity of an "outsider", Jeremy Corbyn, in the Labour Party leadership contest. People are being banned from voting in the leadership election, and having their membership of the Labour Party revoked, because the Party doesn't want Jeremy Corbyn as leader, perceiving his as "too old-fashioned, too left-wing." However, many grass-roots Labour supporters are in favour of Corbyn – a fact that the Daily Mail, in a recent editorial, dismisses as the "...effect of the left-wing echo chamber of Twitter", a statement which neatly glosses over the *right-wing* "echo chamber" of the comments section on their website, and the letters page of their paper. People search for people who share their worldview – this has always been a central fact of human existence; social media hasn't created the "echo chambers" - it's just made them easier to find, easier to access. And, yes, unfortunately it has also made it easier for people to join, or attempt to join, a cause they don't really have a connection with, or passion for, in order to support a suddenly-popular individual who is part of that cause. But, again, that's always been a risk that causes and institutions took – it was just more manageable in the days when we were less obsessed with the cult of personality.

I am grateful, as a natural introvert, for the existence of a means of finding people who think the way I do, whose views I share, and with whom I can have genuine conversation, without being paralysed by anxiety, which, at times, can be so

severe as to render me non-verbal. I'm glad
that, as a person with mental health issues, I
can find, and talk to, other "functionally
dysfunctional" types, rather than being stuck
with the hopelessly unemployable that
psychiatrists think would "benefit from being
around someone like you." (They didn't – it was
an awkward experience for all concerned.) I am
pleased that my wife, who has Asperger's
syndrome, and is an unconventional woman, can
find and talk to other "Aspies", other women who
have looked at the "beauty contest" of
traditional femininity, pulled a face, and
walked away. But I worry that we are losing the
ability, as an increasingly global society, to
interact in a constructive, professional manner
with those who *don't* share our views. Social
media seems designed to throw us into one of two
modes whenever we encounter a dissenting voice –
attack or convert. Neither of those modes
encourages productive conversation. Neither of
them engages the other person. Neither fosters
tolerance and understanding. Instead, the
generations of "digital natives" that are coming
through are being raised with the belief that
conversation is a "battle" that they have to
"win" – that they've "failed" if they don't get
people to agree with them.
   That's not the case. Conversation, dialogue,
in it's highest and best form, is a dance, a
duel. Imagine a fencing match, where the aim is
not to batter your opponent to the floor with
powerhouse blows or vicious kicks, but to
gracefully and delicately find the gaps in their
armour, the chinks in their defences, and dance
with them until they end up in a position where
they admit your superior skill – and where
they've learned something from the process.
Fencing isn't about humiliation, and it doesn't

rely on brute strength. It's beautiful both to
be involved in, and to watch, and no one is
disappointed if a match ends in a draw. Genuine
dialogue allows and encourages people to see the
flaws in their arguments, the gaps in their
knowledge. Much of social media dialogue – and,
indeed, contemporary editorial in print media –
simply calls the other party names, sticks its
tongue out at them, and swaggers off, safe in
the embrace of its own "tribe." That's not
productive. When you call someone a name, you
don't change the way that person thinks – you
merely make them determined to "show you", to
pay you back for humiliating them. When you are
rude, aggressive, and opinionated, you encourage
people to believe all the more in their views,
to defend them more aggressively, and to attack
dissenting views more regularly. Social media
should be s force that draws people together,
and encourages co-operation, debate, and
learning; instead,it risks being the one thing
that divides us.
   As we increasingly teach our children and
young people to rely on and use technology, we
need to take care to also teach them to
distinguish opinion from fact, to have a
conversation with someone who holds different
views, rather than an argument, to express
themselves clearly, and in a way that can't be
misinterpreted. We need to explain – now more
than ever – about the importance of the words we
use, the importance of when and how we chose to
engage. We need to teach constructive criticism
– how to offer it, and how to receive it – and
we need to discuss nuance, shades and layers of
possible meaning.
   We need a "switched on" workforce, yes – but
we also need a workforce that understands the
uses, and potential *ab*uses, of language. We need

people with tech savvy, but also empathy and compassion. We need people who feel passionately about causes, but who can convey that passion clearly, coherently, and gently. We need people who understand that passion should be a fire, that warms all who approach it, and which draws people from miles around, rather than a blunt instrument that batters and silences others. We need to learn to beat our own drum so that it provides an alternative sound to the drums of others – not so that it silences them completely, and deafens those who listen.

Songs, and poetry, were originally a way to convey the stories of the tribe, in a form that was easy to recall, and which could best convey tone, inflection, attitude, enthusiasm, doubt, etc – they encouraged a vibrant, living, interactive form of history, and a form that both required and promoted attention.

The songs of the Internet, by contrast, risk becoming a series of atonal, discordant soundbites, that strip from us the skills of sustaining attention, perceiving subtleties and depths of meaning, and subtly altering and continuing the story in our own voice. Technology, which was hailed as the zenith of inclusion, now risks shutting people out just as surely as the "old world" of communication did.

Recently, a reading group I'm a member of on Facebook had a post from a visually impaired member, asking if people could stop just posting a photo of the cover of whatever book they were talking about, and actually take the time – a minute, at most – to type the title and suthor, as her screen-reader didn't recognise images. There was a flood of "oh, gosh, I didn't realise!" type comments... and, the very next day, people were back to the two-second photo upload from their smart phone. One member even had the cheek to sneer at me, a member of group admin, and another member, for typing in the title and author, asking if we "didn't understand how to upload photos to Facebook", or whether we were "unemployed layabouts who can't afford smart phones." When the admin pointed out the request from the visually impaired member, the comment in response was "well, I'm not going to waste my time accommodating everyone. Perhaps they shouldn't be using social media, if they have a problem with images."

Human beings are a highly visual species – cave drawings offer proof, if it is needed, that we have always been drawn to telling our stories in pictures, rather than with words. It's small wonder that we now expect blog posts, articles, and status updates to come with "photographic evidence", a picture for us to look at, and, ideally, one that saves us actually having to read the whole post, article, or whatever. But that ignores the point of evolution, which is to move beyond, step by step, the limitations of the original, raw material – we will always be a primarily visual species, but our increasing social ability should mean we have better empathy, and more willingness to adapt, to those who are unable to engage with a heavily visual culture. If we can't manage that, what's the point of all this so-called "progress" that technology has facilitated? Yes, a picture paints a thousand words – but only if the artist has sufficient skill, which usually comes from the knowledge and ability to tell the same story in words, so that the picture is a choice, rather than a necessity.

On a more picture-positive note, the rise of emoticons, whilst irritating to some people, is fantastic for those of us with mental health issues or neurodiversities, who sometimes struggle to find the words to describe how we're feeling. I go non-verbal at times, because of schizophrenia. It's not that I'm stupid, or that I don't have sufficient language skills to explain in words how I'm feeling, it's that, temporarily, access to my language centres has been denied. Sometimes, my schizophrenia results in a "shift" to the consciousness and awareness of a ten-year old child - Brae doesn't have the same vocabulary Ashley does, and, as a child, relates better to the bright visuals of "emojis." Even when I'm myself, and as lucid as I ever get, I will often use a smiley face in writing where I would naturally offer one in spoken conversation, and for the same reason - to convey that I intended to be friendly, or lightly sarcastic, rather than (as misinterpretation could easily lead people to believe) critical or dismissive.

Just as technology should be a tool we use, rather than a god we rely on, so pictures should enhance our stories, not seek exclusively to tell them, to the exclusion of language.

In an article on professional networking site LinkedIn, Carlos Munda discusses the role of social media in preventing another financial panic, and subsequent meltdown, in response to the recent Chinese stock market crisis. He describes social media influencers as "modern day cowboys, riding to the rescue and herding the cows away from the precipice." He points out that, when people are able to discuss - and correct - news reports and perceived "facts" in real time, when they have instant access to people who genuinely know what they're talking about, panic can be quickly subdued, and hard facts brought in to throw cold water on the fires of speculation. We are, increasingly, moving away from a reliance on, and trust in, the authority and knowledge of journalists, and towards a point where we engage with one another in an attempt to better understand our world, and the things that are happening in it. In terms of the evolution of human communication, this seems to be taking us on a circular path, back to the way society operated before the rise of skilled, authoritative orators, and, later, the invention of print media, and then of radio and television, and their broadcast capabilities.

 Of course, as another LinkedIn user commented, things could easily have gone the other way, with the "echo chamber" of social media fuelling fear and panic. But that has always been the risk of communication – that the wrong message is spread. We live, and always have done, in the shadow of the knowledge that we can't control the reaction to our words, to our message – all we can do is seek to communicate our position as clearly and concisely as possible, leaving as little room as possible for misinterpretation or manipulation.

GODS

No one is entirely sure how religion came about – not "a specific religion", but the concept of, and need to believe in, gods at all.

We do know that some sort of belief in beings more powerful than humans is as old as human beings themselves. Every culture and tribe that's been encountered has a "creation myth", and, for the most part, a line of similarity can be traced through all of these, however different and disparate they may at first seem.

My personal take on the "why, how?" of the origin of religion is that it springs from what seems to be an innate human trait, that of looking for – and seeing – patterns in things. Religion, gods and goddesses, good and evil, punishment and reward, were ways of harnessing this propensity for identifying patterns, even where none existed: if the harvest failed, it was because of the rains, and the rains had come at that time of year because some invisible, immortal, all-powerful being was angry. Human beings attacked one another, not without reason, but because the gods were engaged in battle themselves, and the energy generated sent men mad. When you were able to get a good price for your crafts, crops or livestock at market, it was because one of the deities had smiled upon you, and whispered to your customers that they should pay you well; you had obviously done something to please her. Human beings have never been good with accepting the randomness of life, or its casual cruelty, and we probably never will be.

And that was why we came to need a new god; the old ones were too unpredictable, too prone to fits of temper, to sulks that lasted for centuries, and to showing favours indiscriminately, to mortals who, in the eyes of their compatriots, had done nothing to merit them. So, we killed our flesh-and-blood gods, our "SuperMen", and came to create a dependable, reliable god that we could predict and control – technology.

"Garbage in, garbage out": if we enter the "prayer" of coding correctly, we're guaranteed our desired outcome. If the outcome doesn't happen, it's because we got the liturgy wrong, not because the Ineffably Divine was throwing its toys out of its pram for unknown reasons. We didn't have to sacrifice our virgin daughters; we could simply go back, find where we went wrong, put it right, and hey presto, our new god delivered.

In his book Blue Like Jazz, author Donald Miller describes his childhood "slot machine God"; "...it doled out rewards based on behaviour, and, perhaps, chance." Human beings dislike chance – it's why we allow ourselves to be fooled by nonsenses like the Gamblers' Fallacy, which states that, inevitably, a losing streak *must* be followed by a run of wins. It's why we talk of "strategies" for winning the board game Monopoly. We simply can't accept that there are things beyond our control.

And thus, as we moved away from the ironclad rule of the Church, as the State became an independent entity, we sought to create our own, dependable gods. We created technology, and called it Lord and Saviour of mankind.

And then, as is inevitable in the saga of humans and gods, we came to fear the god we had created, and, just as the traditional religions used their gods to condone the very worst that human beings were capable of, just as they perverted their holy texts, and brought their gods firmly under human control, so we sought to subdue and control the god we'd created in technology. Tired of worshipping any god at all, we sought to become gods ourselves.

The purpose of gods is control – humanity created them to fulfil its need to control the uncertainties and insecurities that creatures with highly developed social skills and complex social networks are inherently prey to, and we seek to become them in order to end those uncertainties and insecurities once and for all – we first made technology a god because technology promised to erradicate uncertainty, and wrap us in the comforting security of the binary, and we seek to become gods ourselves because, true to its godlike status, and the nature of things we deify, technology revealed truths that terrifed us – unpredictable truths, truths that shattered our illusions of security, or, worse yet, used claims of security to make us feel *insecure.* Just as we hadn't been able to allow the fact that the Christian god we'd created would "punish" us with the loss of jobs, with poverty, with illness, with the inability to find a suitable spouse, with poorly behaved children, or the inability to have children, and so had become gods, and used the philosophy that had informed the creation of the Christian god to inform actions and attitudes that were wholly human in origin, so we found ourselves unable to allow the fact that the god of technology wasn't providing us with a better, more secure and less uncertain future, and decided it was high time we took control of technology, and used its god-like powers to spread a single, unified message – the kind of message that calmed us, that allowed us to put aside our fears, and get back to pursuing our greed, our agendas, and our ambitious paths to glory or destruction.

Gods, of course, need worshippers: without them, the gods, and the religion they promote, fade and die. And we, having made ourselves gods, were no different. At first, we thought we could simply take the worshippers of technology – but that didn't work. The god of technology had allowed and encouraged freedom, and its worshippers sensed that these new gods weren't going to continue that practice. So, we had to tell a story – our own creation myth – that would encourage new people to come and worship us, their new gods.

We told a story of heroes, of people setting out to free information from its chains of corporate control, of people releasing the concept of work from the prison of disappointed expectations and defeated hopes, of people pursuing radical inventions, the cures for diseases we'd barely known before, the final realisation of a life of unbridled leisure, a life where we "worked", at something which engaged, enthused and fulfilled us, for a minimum number of hours, a life in which the remainder of our time could be spent "having fun." We spoke of a more open and transparent democracy, of faceless corporations being held to account by "the little guy", of handing "power to the geeks, respect to the nerds." We fulfilled the seemingly empty promise of the Christian god, that "the meek will inherit the earth."

And people believed us. They came to us, and worshipped us. We had our worshippers. We had become gods.

But the life of a god, we soon discovered, is not a life of leisure. For a start, you have to keep up the stories, and you have to keep inventing new ones. You have to keep an eye on your worshippers, in case they start to see through the thin tissue of your stories, or in case they start to get ideas about becoming gods themselves – there's only so much ambrosia to go around, after all.

And there's always new religions springing up. In the case of we who had taken over god-hood from technology, it was the religions of austerity, the New World Order, and "slow living". People are making themselves gods of less, gods of the natural, gods of fear and uncertainty, and they're telling stories that are more believable than ours. They're starting to seem more powerful – and gods without power die, because gods without power quickly become gods without worshippers.

In the West, we worshipped the gods and goddesses of the Pagan tradition because of the control they gave us over the passing of the year, and the demands of growing sufficient crops to ensure our survival. As things became mechanised, the power of those gods began to wane, and we began to worship the Christian god, who very much resembled the bosses we'd come to know – swift to punish transgressions, but protective of his own as long as suitable tribute was paid. Then, as we moved into a period of massive instability, a period which saw two cataclysmic wars within a generation, that god began to lose his power – the world was uncontrolled and uncertain, the order we loved and needed was bleeding out on so many foreign fields. And so, out of this chaos and disorder, we created a god of the technology that had saved us, first through armaments and their mass production, then, later, through the machines at Bletchley Park, and their remarkable abilities. A god which is losing its power because of the things it has made known to us.

We are becoming the new gods – but we have
forgotten what happened to all the old gods: we
have forgotten that we, too, shall pass.

MACHINES

We set out to make gods of ourselves, and we claim that, through our mastery of technology, we have *become* gods. But have we, in fact, become merely machines? And, if we have, where did we go wrong? How did we, who would have been masters, end up as servants?

The majority of people are aware the "traditional" media outlets, such as television, newspapers, and radio, are owned by companies and organisations with their own agendas, who are committed to putting across a particular point of view. Far fewer people, however, have the same awareness when it comes to the Internet – there remains a widespread belief that the Internet is "free" and "open", that it is owned and run by people who have no agendas or vested interests. True, the recent Public Offering of Facebook, and the purchase of seemingly everything by Google (sorry, Google, but I DON'T want the name I've given a YouTube playlist to be the name that's displayed on any emails I send), may be going some way towards changing that view, but far too many of us still accept, unquestioningly, things we read online – particularly if they come packaged in a handy, shareable meme.

Recently, for example, someone I know on Facebook had shared a "postcard", from the Trades Union Council, via the Independent, that claimed that "workers on zero-hour contracts earned, on average, £300 a week less than a permanent employee." Shocking. Outrageous. There should be protests! Why aren't people rioting in the streets?! But let's look at that statement, shall we?

A 40hr minimum wage job will only pay you £260 a week, before tax and National Insurance are deducted. Most sectors that use zero-hours contracts also only pay the minimum wage to the majority of their staff.

Not the same story now, is it? Now, you're wondering what to be most outraged about - the fact that someone can be employed with no guarantee of actually working (or getting paid) on any given week, or the fact that someone can work for 40hrs a week and earn less than £300 in that time.

It's the same with the hot news at the time of writing - the so-called "migrant crisis", portrayed across almost every media outlet in the heart-wrenching pictures of dead children, washed up beaches as refugees from the Syrian conflict desperately try to get to safety. On social media, particularly, one is almost required to condemn the government's lack of response, to bewail the intolerable racism of Britain, and to generally throw a pity party for people who want pity about as much as they deserve condemnation - that is, not at all. These people are not piteous. They are not helpless and pathetic. They are intelligent, capable, resourceful human beings who want to be able to simply get on with their lives.

What's not so fashionable, of course, is to identify oneself publically with the long-term disabled of the UK, those who will probably always find themselves shut out of meaningful work, shut out of engagement in life, shut out of the realm of other peoples' respect, because of the prejudices that still dog the workplace, because of attitudes that, though they should have died decades ago, but, sadly, live on in people that may shape the hiring processes of companies for another fifteen or twenty years.

The Internet is owned and run, and its memes are provided by, people who want you to think a certain way, because people who predictably think in a certain way are easier to market to – and those who own the Internet are nothing if not businesspeople. They don't want an educated, intelligent population of people who are capable of critical thinking and independent research. They want a population of sheep, who'll willingly buy their products, as long as they're suitably marketed. This ownership by those with something to sell is, of course, the reason for Facebook's controversial "real names" policy – it's not, as Mark Zuckerberg claims, that "...an alias...shows a lack of integrity", so much as it is that aliases are hard to get a handle on. And people who are hard to get a handle on are hard to market to.

As well as being on Facebook, I'm also on LinkedIn, a professionals' networking site; a lot of LinkedIn users are unhappy with the "increasing marketisation" of "their" platform. I'm not among them, because it's obvious that, eventually, *any* Internet site or platform is going to look for a way to make money; most people running Internet sites couldn't care less about the "tribe" that is loyal to the site, or about Internet users more generally. They didn't set that site or platform up out of the goodness of their heart, or because of a personal interest. They saw a market, and, like the businesspeople they are, they went for it, capturing it first with relevant, quickly-liked content, slowly warming that market up for the sell option. LinkedIn was never going to exiat simply for professionals to showcase themselves, their opinions, products, and successes: it was always, eventually, going to be used as a way for its founders to get money – whether by charging users to "promote" their articles and posts, or by allowing advertisers unlimited access to people who, because of the detail they've gone into about their businesses and work lives, are readily classifiable into marketing categories, or through some other means of monetising the fact that they have a vast number of active users, whose details are readily accessible.

On the 27th August 2015, Facebook recorded 1bn unique user logins in one day, for the first time in its history. Now, the venture capitalists who own Facebook aren't simply going to be high-fiving each other at getting 1bn people onto their site (the way my partner and I were when we hit 1000 reaches during an event we recently ran on Facebook, with a registered engagement rate of 20%). No. Those investors are going to be looking at that figure – 1bn – and seeing 1bn people who, in the vast majority of cases, have shared an awful lot of information about themselves, and thus are so easy to sell things to, it's like shooting fish in a barrel. It's why Facebook stubbornly refuses to make it easier for business pages to engage with "real people" - they don't *want* businesses getting a foothold on Facebook, because businesses know what the game is, and they don't fall easily for marketing "spin."

We delude ourselves if we believe that our access to, and knowledge of, technology has made us "more sophisticated" – the people and businesses who see us as nothing more than a "market" for them to make money from have always been several steps ahead of us in terms of their understanding of technology, and its potential.

An opinion piece in my regional paper recently was discussing how few people walk anywhere anymore, and pointed out that, while disdaining this completely free activity, we will spend hundreds of pounds on gym memberships we hardly ever use. Probably still more of us have Wii-Fits, home exercise equipment, sophisticated "step counters" and apps to track our heart rate, calorie intake, and likely age at death. We Google exotic excersie regimes, and complicated, faddish diets. We click through pictures of "desirable" bodies, and fixate on blogs about "body positivity". We don't walk anywhere because technology has convinced us that there's a much better, much cooler, much less reminiscent of poor people and their lives way of getting, and staying, in shape. We don't walk anywhere because technology offers us not only an unlimited range of exercise possibilities – that, of course, we have to study in-depth before we can even think about embarking upon them – but also tells us, in many different ways, that "the way we are is perfectly okay." If we feel bad about ourselves at 9am on a Sunday morning, we can put up a Facebook status, secure in the knowledge that, by midday, we'll have a whole crowd of people rushing to reassure us. (From a safe, non-engaged distance, of course.)

The problem with technology is, it makes it easier to lie – and it makes our lies more believable. More people would lie in a phone call than face-to-face, and still more would lie in an email – and email is rapidly becoming our primary method of communication. Social media likewise facilitates, and, some would say, actively encourages lying; a recent meme on Facebook had the caption; "IF ONLY YOUR LIFE WERE AS INTERESTING AS YOU PRETEND IT IS ON FACEBOOK" - ironic, given the platform on which the meme was circulating, but also true - we have to keep reminding ourselves, and *being* reminded, that people edit the lives they present on Facebook – they hide certain aspects, embellish others. They twist the truth, and they tell outright lies. A lot of column inches and hand-wringing has been given over to *why* people do this. My own theory, in contrast, is relatively short and (not so) sweet; social media allows us, for the first time in human history, to truly realise the vastnass of the world, and the wealth of opportunities open to us – and, terrified, we fear that we are insignificant in the face of all of this. And so we create significance for ourselves, a significance devoid of meaning, but which will readily impress others.

Jeremy Corbyn – a supposed "rank outsider" - was recently elected leader of the Labour Party in the UK, after a dirty-tricks campaign on social media and in print media against him – not just by the Conservative Party, but by other members of the Labour Party. Facebook, in particular, was up in arms about memberships for the Labour Party being revoked, and applications for membership prior to the leadership election being denied; I could never say this on Facebook (the conspiracy of silence the machines have us trapped in,) but the reason why is obvious; people were openly supporting and promoting one particular candidate, and openly declaring their intention to vote for that candidate, across a broad public forum, in what was intended to be a hidden ballot, as all UK elections are.

Will Jeremy Corbyn make a good Labour leader? Would he make a good Prime Minister? Author Robert Kelsey seems clear on the fact that he won't, but, then if Corbyn *does* succeed, Kelsey will have to eat his (several books' worth of) words on how "genuine outsiders can never achieve success within the traditional, insider-run system."

Jeremy Corbyn seems a lot like me – someone who's taken on the world and lost a few too many times, someone who's tired of fighting, yet who has no option but to carry on, someone who's managed, one way or another, to lose pretty much any friends he had. A definite outsider. A loner. An introvert. Crazy. Dangerous. A guaranteed failure. And yet, I know from experience that I *can* succeed, with or without the support of others, if I'm passionate enough about something. And I sense a passion for reform of Britain in Corbyn – like calls to like, and I recognise his song in the echoes of my own.

WHERE ARE WE GOING?

We are, it seems, heading towards a point where technology is no longer just "part of" our daily lives, but IS our daily life. Perhaps we're not just "headed" there - perhaps we've already arrived?

Think about it; whereas a typical morning routine might once have been simply: Get up, have a shower, have a cup of coffee whilst reading the paper, get dressed, now it is almost certain to include at least one, if not more, of the following: Check emails, update social media platforms, scan news websites. I'm sure you can all think of other, tech-related tasks that have been added to the first-thing-in-the-morning "to do" list. Where we used to worry simply about being uninformed, now we worry about being, or appearing to be "out of the loop." We worry that our social media profiles portray as us being too serious, or not serious enough. We tell "white lies", we exaggerate facts, and, as long as there are pictures to back them up, we don't question other peoples' "facts", even though, in the twenty-first century, with PhotoShop and airbrushing, and a plethora of screen-ready images just a click away, the camera usually lies.

Jobs are lost on a daily basis to automated processes, while offices regularly grind to a halt whenever the internet goes down for more than five minutes. Rural businesses complain that their broadband speed isn't fast enough, while urban businesses whine about the lack of skilled coders, the derth of tech-savvy, work-ready employees. Those providing content for the endless pages spawned by the business end of technology are expected to work for free, while the website owners expect multi-million dollar handouts for a "concept."

At the same time, technology has allowed greater participation in the political sphere than ever before. It has made it easier for people to find jobs, and to set up their own businesses. It has revolutionised the way businesses and ideas are funded, and who is able to access that funding. It has made steps towards giving power back to the people; we're a long way from being there, yet, but a start has been made, and things that are started have a way of propelling themselves to a conclusion.

It will most probably be technology, not quantitative easing or government intervention, that stabilises existing economies, and establishes others – as more people are able to set up in business, source funding for passions and ideas, undertake courses and attend seminars, as well as find jobs, through technology, so more people will be in work, and the demand for State support will fall. At the moment, we're in a transitional phase, where the traditional workplace and its practices, the traditional ways of establishing and funding a business, and the traditional methods of finding a job, are falling away. Transitional phases are ususally messy, usually expensive – that's what we're seeing now. Sometimes, people die before the transitional phase comes to an end. It's unfortunate, but it happens. We've been here before, and we'll be here again, gazing into the abyss.

As people come to realise that they can have pretty much the entire world if they have a laptop and a broadband connection, so the attitude of "bigger and better", the demand for "more stuff" will fall away – we're already seeing a surge in interest in spirituality and "alternative" ways of living and being, and this is at least in part fuelled by technology, which allows easy-access to the lives and thoughts of others through blogs, social media profiles, and YouTube videos. We are easily able to get hold of as much information, from as many different points of view, as we want, and have it presented in a way that is easy for us to understand and engage with. Even if we're only looking into "traditional" religion, we're not any longer stuck with the church down the road – we can log on to the Church of Fools, we can watch podcasts from Jews, Muslims, Christians, Sikhs, Hindis and Ba'hai who are just like us. We can attend a service at 3am on a Monday morning, or 3pm Thursday afternoon. We can even create and promote our own religion – and probably find others ready to engage with it.

We can change our name as easily as changing our image, find a partner with as little effort as it takes to find a kitten, and explore several conflicting attitudes and ideas before lunch. We can write books without being authors, produce records without being musicians or singers, make films without being actors.

Yes, there are downsides, disadvantages, and outright dangers surrounding the potential of technology, but that's the nature of transitional phases – it's why not everyone makes it out alive, whether the transition is from slavery in Egypt to the "promised land" of Canaan, or from unemployed and unemployable disenfranchised to global thought-shaper.

The challenge, for us, is to understand, not the technology itself, but the fact that the technology *isn't going awsy*. We can afford to start being serious with it, now, because there will always be time for us to play with it. Like the products and processes of the Industrial Revolution, the impacts and effects of the machine age will be with us long after our

lifetimes have ended. And, like the Industrial
Revolution, the machine age will change the way
we live, work, and play forever. It will change
the value we place on "childhood", and the
meaning of "leisure." It will come to shape not
just our lives, but our landscapes.
 It took almost two hundred years for the
Industrial Revolution to bring wealth to the
people who performed the processes, rather than
simply those who *owned* the processes. And, as we
come to the end of the industrial age, we're
still only just there. Technology moves faster,
and changes happen in a shorter time span, but
technology, for all its speed, is still in very
relative infancy in terms of being a new
paradigm; the issues around exploitation of
content providers, copyright,income inequality,
etc, will be resolved – we simply need to wait,
and re-learn the patience we lost in our wonder
at technolgy.

 Charlie Brooker's series of films for Channel
4, Black Mirror, terrified many people with
their suggestion that we could "live on" after
our deaths through our interactions with
technology, and, indeed, it's entirely possible
that, fifty or a hundred years fron now,
computers will be powerful enough, and robotics
advanced enough, that a "human identity" could
be constructed from social media postings,
YouTube videos, and the cookie-crumb trails of
website use, and transplanted to a robot form.
It is possible that, one day, technology will be
able to render us all as holograms, and to
operate multiple versions of "us" with minimal
input from the "real self." After all, the idea
of a "digital afterlife" isn't so hard to
imagine. It's not even particularly futuristic –
think of how much we feel we know about

Shakespeare, Emily Dickinson, Richard the Third, and many others who lived and died long before the rise of technology. They live on in their writings; a digital afterlife would be no different to scholars reconstructing handwritten texts, except for the fact that the reconstruction will be done by computers, from the strings of code that make up our digital lives.

Digital lives are already a reality – for some people, more of one than their physical lives. Business is carried out on SecondLife every day, meetings and interviews take place over Skype, we look at someone's LinkedIn profile rather than their CV. People come to believe that they are the selves they present online, to the extent that they end up bewildered and uncomprehending when they are accused of "deception" after continuing those digital selves into the physical world.

My partner enjoys playing SecondLife – five years younger than me, and from a relatively privileged middle-class background, compared to my working-class roots, she is very much a digital native. When I tried to get into SecondLife, I ended up lost, alone, and facing a brick wall – as I joked at the time, the exact replication of my *first* life. For Morgana, technology allows for a re-imagining, rather than a re-invention; she is balanced enough to have interests and enjoyments that are firmly grounded in the physical world, such as knitting and jewellery making, to run alongside her digital dalliances. But we can both see the allure of a world without limitations, a world where, as long as you press the right buttons, take the right actions, you'll get where you want to be. The physical world, frustratingly, doesn't work like that, no matter how much the self-help industry may claim it *does*, if you just "understand the secret" (and no matter how much money they make selling "the secret" to the gullible.)

Technology hss the potential to grant us unlimited power over our lives – to make us, in effect, the gods of our own destiny. But, with great power comes great responsibility, and this, it seems, is something we have yet to come fully to terms with.

We're going back to how we first began – with the ability to exercise a great degree of power over our lives, and the co-dependent responsibility to ensure we didn't abuse that power by not working sufficiently to provide for our families, or by exhausting a particular food source, or damaging a means of food production.

Those same responsibilities, though framed somewhat differently, still lie heavy upon us as we engage with the machine age, the technological age, and the return of personal power it will bring us.

A business can be set up in under an hour – it may never become profitable, but it can exist. A book can be published in under a week. Secrets can be shared – and spilled – in seconds. Lives and reputations can be ruined in an instant.

I talked, in the previous chapter, about the "dirty tricks" campaign, aided and abetted by the media – mainly print media, but social media certainly got a look in – run against Jeremy Corbyn, the new Labour Party leader. Just days into his leadership, and the media are already spinning stories, taking things out of context, and rushing around in an effort to discredit him.

I don't want to live in a world where who vote for, what I believe, my attitudes and behaviour, are things engineered by distant, uninterested third parties using marketing and manipulation in increasingly sophisticated ways. I do not want to find myself, in 2024, living George Orwell's *1984*.

We already have to put up with a world in which businesses guard their secrets better than the crown jewels, and yet, at the same time, those same businesses think it is perfectly acceptable to invade peoples' privacy if those people have been foolish enough to seek employment. We already live in a world where – despite legislation supposedly guaranteeing the "right to a private life", whether or not an individual has had an abortion, surgery, or gender reassignment is considered "in the public interest" by media outlets that will happily spew those facts anywhere and everywhere. (Heads up, ITV News – It. Is. Not. Relevant that someome convicted of a crime – ANY crime - "was born a woman/man, and used to be known by XYZ name." Would you point out that they were of "an ethnically distinct minority?" No, you wouldn't. I know other places do this – ITV were unfortunate enough to cross my radar.)

I don't want to live in a world where people think that "because it's just online", it's okay to mnake other peoples' lives a misery, but nor do I want to live in a world where everything is trigger-warninged to hell, where people lie in wait for someone with assumed "privilege" to "slip up" and make a remark that falls foul of some kind of -phobia or -ism. (As a white male, I'm tired of having to "out" myself as having undergone gender reassignment in order to be allowed a voice in online spaces. I'm tired of the assumption of privilege, the assumption that, because of a life people imagine I've had, because of facts of biology and upbringing I didn't choose, and can't change, nothing I say will have any value. Oh, and I'm tired of the put-down of "Oh, bless – you don't like it when your voice isn't the loudest in the room, do you?" My voice has rarely, if ever, been the loudest in the room – I learned early, and learned hard, not to talk, not to have opinions. I'm still getting over that.)

Technology was meant to make things more open, more transparent – instead, there's a very real possibility that we will end up living in a world where no one says what they mean, where some people aren't allowed to speak at all, or, at the very least, are too afraid to, and where the truth of someone's words is judged by their ethnicity, their socioeconomic background (actual or assumed) and their gender – there's already been an instance, roundly condemned on social media without any evaluation of the facts, of a white male poet submitting a work under a female Chinese name.

The pertinent point? That the poem – which had, apparently, been rejected when submitted under the poet's own name – was published. The exact same poem. The selectors, if this is true, didn't even *read* the poem – they read the name of the poet.

I don't want to be shut out of the world because of something it isn't in my power to change. I don't want to be silenced, because, rather than hearing the truth in my voice, all people hear is the noise of voices of people like me, howling through the centuries.

Your experience isn't valid, because you're a n*****. Your experience isn't valid because – privilege! Spot the difference? I want technology to make it easier to open up dialogue, between anyone and everyone – I don't want it to be used to shut people down, and yet I fear it will be.

The issue is, the Internet, by its very

nature, creates and facilitates arguments. It polarises, it draws attention to points that can be misinterpreted – sometimes willfully – and it encourages the "Us vs. Them" mentality – after all, everyone wants their posts to be popular, and what's more popular than a vicious, heated argument where people can easily take sides and have their say, even if the issue under debate is nothing to do with them? All the complexity of the information age boils down to the simplest, lowest-possible denominator with the Internet.

   That's why "privilege" has become the big deal – because the Internet doesn't allow for the complexity of the fact that the REAL issue, the real problem, isn't privilege, which can't be helped, but **entitlement**, which can be helped. The Internet doesn't allow for the point that women and ethnic minorities can be entitled, that people who are self-educated can be entitled, and that, yes, white, able-bodied, middle-class, heterosexual males can be entitled.

   Privilege is the position society confers on you, something given automatically, and that you can't renounce easily. Privilege is whether you're deemed a god or a machine.

   Entitlement is how you use or abuse that privilege – do you use it to lift up those without privilege, or to berate, denigrate, and crush them? When someone uses "inappropriate" language, do you explain to them why their words are problematic, or do you just blast them and their words across your own social media channels, and call them "privileged" and "offensive"? Educational privilege, for example, isn't just restricted to having a degree, or having been to a good school – it also encompaases being able to attend discussions and debates, having access to, and understanding of, current reasearch and opinions on various issues, living in the place where the changes to language, etc are made.

Because language *does* change. And not every one is made aware of the changes at the same time.

Two recent examples of change of language that I hadn't been fully aware of: it is no longer considered polite to describe someone as "in a wheelchair"; they "use a wheelchair." Fine, fair enough, I will try and remember this. But, I only know about it because I have the privilege of being able to read, procure, and afford The Sunday Times, which carried an article mentioning this fact. Without that article? I wouldn't have realised there was anything wrong with saying someone was "in a wheelchair" – until I got yelled at about it on social media. Because the people who live in the area I live in, the people I spend most of my time with, still use "in a wheelchair" – without any negative intent whatsoever. It's a neutral statement.

Then, a couple of days ago, there was what I felt to be a wholly inappropriate "blast" (where someone with a large social media following screen-shoots and shares a conversation/post from someone else, usually someone they've taken issue with) where a trans* man was slated for posting, on his own timeline, that he "only dates bio guys." Bio guys, for those not in the know, was a phrase that was in current use when I transitioned nearly a decade ago, and simply refers to men who were born male-bodied. The blast happened because a non-binary individual objected to the use of "bio guys" – yes, it IS now an old, and mostly redundant, term, yes, it would have been much easier, and possibly (only possibly – social media seems to attract the kind of people who like to have problems with what someone else said, what they meant, etc) more polite, for this guy to simply say he only dated trans* men. But he didn't; he used language as he understood it, to convey a personal preference. The non-binary individual communicated their displeasure quite vehemently (I've yet to meet anyone who describes themselves as "an activist" who is capable of politely and privately drawing attention to issues they have with someone else), and got blocked by the OP (Original Poster.) As, to my mind, they should have done.

Now, I was aware that most people don't use "bio guy/bio woman" any longer, and I can see why. However, I find the neoism that's replaced it – cis – as offensive as most activists find "bio guy/bio woman." If I wish to make it clear that someone is living and presenting in a gender which matches the biological label that was stuck on them at birth, I will simply say that person is "non-trans*". Cis, to my mind, conveys a level of contempt and dismissal that I will not countenance – most of my friends are non-trans*; it doesn't mean they haven't gone through some pretty rough stuff in their lives, it doesn't mean they don't care, it doesn't mean they are entitled morons. And I refuse to use language that implies any of that.

But the Internet doesn't allow for individuality – we live in a global world, but one that's becoming, if anything, *more* tribal. The Internet allows us to be fiercely protective of our opinions, because we don't actually have to see the look in another person's eyes as we assert them. We don't have to risk getting a quite literal slap, or being told, in front of a roomful of people, that we're "a f****** prat" - and being told that on social media somehow doesn't have quite the same effect.

I've come across a couple of posts where people have admitted that they "didn't realise" how serious a social media attack on someone, that included aggressive posting to business pages, etc, was - that it "was only the Internet - it didn't really occur to me that I could actually cause anyone a problem in the real world."

In both cases, the posters were so-called "digital natives", people who have grown up with the Internet, people for whom technology is second nature. And yet, for whom it still isn't seen as part of the "real world"; it's just another computer game, where everything's pretend, everyone's having fun, and no one ever really gets hurt.

This stems, of course, from an obvious disconnect - technology has granted individuals an immense amount of power over one another, and, in many cases, even over organisations, companies, and governments - but the message individuals receive is overwhelmingly "You are insignificant and powerless." Of course it's not going to occur to one twenty-something sitting in their bedroom, seething as they type, that they will have any impact on anything - they've spent most of their lives being told nothing they do makes any difference, that they're just worthless cogs in the machine. No one warned them that technology had the power to make them a god, with all of a god's responsibilities.

As technology becomes more widespread, as more people use it with as little thought as they give to breathing, we are moving towards a place where individuals without credentials, qualifications, or significant experience can wield immense power – we are moving towards a point where institutions and accreditations are falling by the wayside, swayed by the power of the autodidact, the citizen journalist, the social justice warrior, the self-employed 22yr old, the online investor with a healthy passive income. And yet, we still live in a world where people are told, day in, day out, that they are powerless and insignificant.

IS THIS THE AGE OF THE MACHINES?

People have always feared machines, usually
because, when they came, they disrupted,
distrubed and destroyed a known, familiar way of
life, and left people who did not have the
education or the money to become familiar with
them behind. Much of science-fiction writing
revolves around the idea that machines are
intrinsically evil, and that it's just a matter
of time before they become at least as
intelligent as human beings, if not more so, and
seek to take over.

Some might argue that that time has already
come, and point to the apparently vast numbers
of Western children who have never played
outside, who lack imagination, and who give up
on a task if the information required to
complete it isn't available within 5 minutes (or
less) of Googling. The majority of people
responding to surveys say they will navigate
away from a website or page that doesn't load
within three seconds.

News reporting can be instant, surgical
procedures can be seen as they happen, and
distorted facts can be halfway around the world
while the truth is still being typed up into a
Press release.

Our boss can call us on holiday, or at 10pm;
we can lie to someone with a few words and a
click. Images are often taken at face value,
even though they've been altered to the point
where the original would be unrecognisable. We
are at the height of our powers, drunk with the
heady sense that technology enables us to be and
to do anything we want. Millenials – those born
after around 1980 – no longer see the machines
as evil; they see the machines as the means of
becoming gods.

But, in our exuberance, we've forgotten one
thing: gods always die.

BIBLIOGRAPHY

In no particular order, these sources – ancient and modern! - proved helpful in writing GODS AND MACHINES:

TRUST AGENTS, Brogan, Smith, 2010 (John Wiley & Sons)

ALTER-EGOS, Cooper, Dibbell, 2007 (Chris Boot)

THE BINARY REVOLUTION, Barrett, 2006 (Weinfield & Nicholson Reference)

THE BLETCHLEY GIRLS, Dunlop, 2015 (Hodder & Stoughton)

VIRUS OF THE MIND, Brodie, 2005 (Hay House)

NETYMOLOGY, Chatfield, 2013 (Quercus)

THE INFORMATION, Gleick, 2012 (Fourth Estate)

THE WINTER OF OUR DISCONNECT, Maushart, 2011 (Profile)

DARK NET, Bartlett, 2015 (Windmill)

And websites too numerous to mention, but including the usual suspects – Facebook, Tumblr, Twitter, LinkedIn, Google, Amazon, Blogger, WordPress, DirectGov, TheInternetSociety.com.

To anyone I've left out – please feel free to contact me so I can offer a proper citation.

To the numerous folk across social media who got involved, answered questions, made me laugh, and exhausted my patience at times – keep doing...whatever it is you do.

This is it. Our closing lines. The end.